RADIATION THREATS AND YOUR SAFETY

A GUIDE TO PREPARATION AND RESPONSE FOR PROFESSIONALS AND COMMUNITY

RADIATION THREATS AND YOUR SAFETY

A GUIDE TO PREPARATION AND RESPONSE FOR PROFESSIONALS AND COMMUNITY

ARMIN ANSARI

CRC Press is an imprint of the
Taylor & Francis Group an **informa** business
A CHAPMAN & HALL BOOK

Chapman & Hall/CRC
Taylor & Francis Group
6000 Broken Sound Parkway NW, Suite 300
Boca Raton, FL 33487-2742

Printed in the United States of America on acid-free paper
10 9 8 7 6 5 4 3 2 1

International Standard Book Number: 978-1-4200-8361-3 (Hardback)

Library of Congress Cataloging-in-Publication Data

Ansari, Armin.
Radiation threats and your safety : a guide to preparation and response for professionals and community / Armin Ansari.
p. ; cm.
Includes bibliographical references and index.
ISBN 978-1-4200-8361-3 (hardcover : alk. paper)
1. Ionizing radiation--Accidents. 2. Ionizing radiation--Safety measures. 3. Emergency management. I. Title.
[DNLM: 1. Radiation Injuries--prevention & control. 2. Disaster Planning. 3. Disasters. 4. Radioactive Hazard Release--prevention & control. 5. Radiologic Health. WN 650 A617r 2010]

RA569.A56 2010
362.196'9897--dc22 2009030137

Visit the Taylor & Francis Web site at
http://www.taylorandfrancis.com

and the CRC Press Web site at
http://www.crcpress.com

Dedication

To community members and professionals who go the extra mile

CONTENTS

Part Four ~ CONCLUSION

PREFACE

SEVERAL YEARS AGO, BEFORE THE ADVENT of Internet shopping, I bought an old issue of *Life* magazine from a shopping mall kiosk in Knoxville, Tennessee. It was the September 15, 1961, issue and had a picture of a man in a "civilian fallout suit" on the cover with the title "How You Can Survive Fallout." I browsed through the magazine thinking that it represented an era that we had moved beyond. I had no idea that the events occurring exactly 40 years after publication of this particular issue would launch another era during which, once again, we would be concerned about this possibility.

Also, I had no idea at the time that I would be spending the second half of my career working on nuclear and radiological emergency preparedness and response issues, part of which entails lecturing and conducting training workshops across the country. It was through these interactions with a broad spectrum of professionals and community members that I realized that an information gap had not been addressed, even though an enormous amount of information was available.

There has been an explosion of information on the subject of radiation emergencies and related issues in the scientific and technical literature. Multitudes of radiation textbooks ably serve professionals in their respective medical and technical fields. State, federal, and international agencies have created numerous guidance and planning documents, and some of that information keeps changing as new plans are drawn and new terminologies and acronyms are created. Also, numerous fact sheets and information pages on radiation and radioactivity, radiation drugs, and other emergency response issues can be found on governmental and nongovernmental Web sites. Finally, an aspect of commercialism offers information to private citizens as well as government consumers.

Although this vast body of information is valuable and, for the most part, serves the intended purpose, much of it is too detailed or too technical, tailored for specific audiences, or simply too dispersed. Consumers of information who are new to the radiation arena have to sift through a lot of material to find what is important or applicable to them. They are likely to be overwhelmed with information or not get enough of what they need.

My goal in writing this book is to bring together, in a concise way, essential, need-to-know, and practical information about radiation threats in an approachable form and content. The book is written for discerning members of the general public who do not aspire to become radiation experts, but they desire more than just a superficial knowledge of the subject; those who need to understand the "why" so that they can use and apply the information. Resiliency of the public at large is dependent on what they know about radiation and how well they can put information and instructions from emergency response and public health authorities into context.

The book is also written for professionals in various fields of expertise who are new to the subject of radiation and who may be called upon to serve in a radiation emergency and apply their specific knowledge and skills under those circumstances. These include emergency management and emergency response professionals, hospital staff, and environmental health, mental health, and other public health professionals. These professionals need to understand the radiation threat beyond what they may learn from a generic "all hazards" or weapons of mass destruction training.

Last, but certainly not least, this book's coverage recognizes that, regardless of our professional backgrounds, concerns for our families' well-being as well as our own safety will affect our response to a radiation emergency, how well we do our jobs, and how we help our neighbors and communities.

I knew that writing a book such as this for a broad audience would be challenging. I hope that I achieved some measure of success in providing this information in a way that is helpful for the intended audience. This book is not intended to teach basic radiation science and theory to students of the field. However, it can be used in a public health or general science curriculum, or portions of it can be used to develop radiation awareness training for specific audiences. All of it is written in a suitable format for self-study.

I need to thank many individuals. Luna Han, my senior editor at Taylor & Francis, shared the vision for this book from the beginning, guided me in the process of preparing the proposal, and provided valuable feedback throughout. I thank Jill Jurgensen and Judith Simon at Taylor & Francis for their assistance in the final editing and production process. My thanks to Charles Miller, Robert Whitcomb, and my management at the Centers for Disease Control and Prevention (CDC) for their support. I undertook this task, however, in my private capacity and outside my official CDC duties. Therefore, I bear sole responsibility for the material in this book. Any opinions expressed are mine alone (and not those of the CDC). I also have no direct financial interest in any product mentioned in the book.

I have benefited from collaboration and discussions with many friends and colleagues in various organizations. I especially thank my colleagues who took time to review various portions of the manuscript and provide valuable comments. In particular, I am indebted to Anthony Moulton, Amy J. Guinn, and Jeffrey Nemhauser, whose input certainly made this book much better than it would have been without their generous help.

I am immensely grateful to my wife, Elham, who encouraged me constantly, provided me with quiet weekends, served as my sounding board, reviewed early drafts of each chapter, chased down owners of photographs, and, most important of all, handled the lion's share of parenting while the book was written. I could not have completed this work without her. I also thank my parents, Mohammad and Anice, who instilled in me the love of learning and showed me the importance of going the extra mile. Finally, I thank my children, Armon and Kiana, who were so patient with their constantly working dad. I hope that their generation and others to follow will not have to be concerned with radiation threats, but instead will continue to harness the amazing power and promise of nuclear radiation for the betterment of their lives.

Armin Ansari

THE AUTHOR

Armin J. Ansari is a health physicist at the Radiation Studies Branch, Division of Environmental Hazards and Health Effects, National Center for Environmental Health, Centers for Disease Control and Prevention (CDC) in Atlanta, Georgia. He serves as subject matter expert in CDC's radiation emergency preparedness and response activities and has frequently represented the CDC on the Federal Advisory Team for Environment, Food, and Health. Dr. Ansari chairs an interagency working group on population monitoring in radiation emergencies and was the lead author of CDC guidance on population monitoring for local and state public health planners. He serves on a Homeland Security Council interagency committee for preparedness and response to radiological and nuclear threats and was a contributing author to the federal planning guidance for response to a nuclear detonation.

Dr. Ansari conducts training workshops and lectures extensively on the topic of radiological/nuclear emergencies for technical and nontechnical audiences. He was nominated for the CDC's Charles C. Shepard Science Award in 2006. Prior to joining the CDC, Dr. Ansari was a senior scientist with the radiological consulting firm of Auxier & Associates in Knoxville, Tennessee, and a project leader with the Environmental Survey and Site Assessment Program at Oak Ridge Institute for Science and Education.

Dr. Ansari received his BS and PhD degrees in radiation biophysics from the University of Kansas and completed his postdoctoral training at Oak Ridge National Laboratory's Biology Division and Los Alamos National Laboratory's Life Sciences Division, where he was an Alexander Hollander Distinguished Postdoctoral Fellow. He is a diplomat of the American Academy of Health Physics, a member of the Georgia East Metro Medical Reserve Corps (MRC), and a member of the Gwinnett County Community Emergency Response Team (CERT).

Part One

UNDERSTANDING RADIATION

1

INTRODUCTION

THE USE OF RADIATION AND RADIOACTIVITY in science, medicine, energy, industry, and agriculture has brought tremendous benefits to our society. Yet, we live in a time when the threat of nuclear terrorism and the possibility of facing a radiation emergency, including a catastrophic nuclear detonation in one of our cities, are unfortunately very real. In the big picture, the most effective method to deal with these threats is to keep them from ever occurring in the first place. Whether this goal is accomplished through stringent safeguards and security measures for nuclear materials, advanced surveillance and interdiction methods, engaged diplomacy, or all of these efforts is beyond the scope of this book and the expertise of the author. Here, we address what you should know ahead of time, what you can do to prepare, and how you should respond to a nuclear or radiological incident as professionals and members of our communities.

If faced with a radiation threat, regardless of where you live or what you do for a living, everyone shares some common concerns about the immediate safety of their families and themselves, as well as that of neighbors and co-workers. Whether you are an emergency manager or an ambulance driver, a school teacher or a restaurant owner, a hematologist or a surgeon, a hotel manager or a mortician, or drive a bus or a waste collection vehicle, the way you do your job will be affected during a radiation emergency. People also have concerns about long-term impacts that such an incident may have on their families' health, their jobs, and their communities.

The underlying message in this book is that there is no reason why you should feel helpless when faced with a radiation emergency. You can take certain actions to protect yourself and your family. You can take actions to help your community if it is directly affected or help a displaced population from far away communities who have come to you for help. How we react to a radiation emergency will determine its true final impact.

Preparedness is a key component of resilience and the most effective method to reduce the potential impact of such disasters. Just like any threat, a prerequisite for preparedness is an understanding of the nature of the threat itself. Radiation and radioactivity invoke an immediate sense of anxiety and fear in most people, largely due to lack of information. Indeed, most people would have difficulty correctly defining radiation. This inability is not surprising, given that public perception of radiation and radioactivity, at least for many, is formed by images of mushroom clouds, cinema and television dramatizations, and even cartoon characters.

The need for understanding radiation is not only relevant to a radiation emergency that may occur sometime in the future. People deal with radiation issues on a continual basis. Examples include the increasing need for alternative sources of energy, such as nuclear energy, and the dramatic increase in use of radiation in diagnostic medical procedures over the last two decades. Consumers need to be better informed when it comes to using these medical services and to understanding the issues affecting their future.

The main purpose of this book is to provide essential and practical need-to-know information that professionals and discerning members of the general public can use and apply in a radiation emergency as well as in understanding and appreciating everyday radiation-related issues. Keeping in mind this broad audience, all of whom need information they can use and none of whom has any desire to become a radiation expert, every effort was made so that the text

- provides only information that has practical utility or is necessary to appreciate and understand basic concepts;
- uses everyday language that is easy to follow and understand;
- avoids detailed theoretical or technical discussion unless it is absolutely necessary; and
- avoids using equations to illustrate concepts—instead, explaining principles using words and examples.

Most chapters follow a similar structure. They begin with an overview section summarizing and putting in context the rest of the chapter. The main section of each chapter has detailed discussion in plain language with descriptions and examples as appropriate. The text is divided into multiple headings so that readers can easily navigate throughout the chapter. As much as an author may desire otherwise, many readers will not read the text in sequence and they may not be interested in every topic. The many cross-references will facilitate navigation for these readers.

Most chapters include one or more brief paragraphs starting with the phrase, "Did you know?" These segments are not essential information, but rather are included to provide additional context, put information in perspective, or simply offer nuggets of information readers may find interesting.

Each chapter ends with a short, annotated list of selected resources for readers who need additional, in-depth information.

Finally, it is common to use only SI units in publications.* Although most of the world has adopted the international units of radiation *in practice,* the fact remains that in the United States, federal and state government documents, first responder training courses and manuals, and radiation detection instruments still use the conventional units of radiation, and the practice is likely to continue for the foreseeable future. For this reason, conventional units are listed in parentheses each time the international units are used.

* The International System of Units (SI) (see http://www.bipm.org/en/si/).

2

RADIATION IN EVERYDAY LIFE

OVERVIEW

Did you know that every person in the world is exposed to radiation every minute of every day and night? When people learn this fact, they may think initially of microwave ovens or cellular phones as sources of radiation, and they would be correct. However, let us consider a different type of radiation, one that has the ability to penetrate and rip apart atoms in our bodies. This is "ionizing radiation" and people are exposed to it constantly, no matter where they are or what they do. In fact, ionizing radiation is as old as the universe—billions of years older than microwave ovens, cellular phones, and human beings.

The sources of this type of radiation on Earth are numerous. Ionizing radiation comes from outer space, from radioactive atoms in the soil beneath your feet, from radioactive gas that emanates from soil and rocks and seeps through your home, and from radioactive atoms buried deep in your body tissues. If ionizing radiation, in fact, is so powerful and pervasive, why are human beings not badly mutated, dead, or extinct?

Ionizing radiation has been a part of the natural environment since the very beginning of Earth and is as natural to Earth as is oxygen. All living species, including humans, have evolved mechanisms to cope with and live in this "sea of radiation." It is significant that, as far as is known neither humans nor any other species have evolved any sensory capability to detect ionizing radiation. This strongly suggests that there was no evolutionary need to develop that capability. In other words, people's bodies are accustomed to certain natural levels of radiation and radioactivity and can deal with them successfully.

On the other hand, radiation in excessive amounts can pose serious threats to human health and can even kill. The key issue is the amount or the dose of radiation. At very low doses, the risk is so small that it cannot be measured. As the radiation dose gets higher, however, the probability of harm becomes greater.

The main purpose of this chapter is to provide a basic understanding of the natural radioactivity in our lives. This understanding provides a baseline against which you can compare any additional sources or doses of radiation and evaluate the threats they may pose in a meaningful way. We will illustrate this by means of an example, but first must make a note about units of radiation.

Some unit of measurement must be used to indicate amounts and doses of radiation. Chapter 3 covers this topic in detail, but here we will use one unit to make a point about natural or baseline radiation. The international community uses a radiation dose unit called the millisievert, which is pronounced "milli-see-vert" and abbreviated as "mSv." In the United States, the older unit of the millirem, abbreviated as "mrem" (1 mSv = 100 mrem), is typically used. We will use both units in this chapter.

Now, back to the example: How would you gauge the relative hazard of a 0.01 mSv (1 mrem) dose of radiation? How dangerous is that dose of radiation? What if you knew that on a typical commercial flight from Washington, D.C., to Los Angeles, each passenger receives approximately 0.02 mSv (2 mrem) of radiation from space? What if you knew that, on average, every person is exposed to 3 mSv (300 mrem) of natural background radiation every year simply by virtue of living on Earth? Given these facts, how would you rate the relative hazard of a 0.01 mSv (1 mrem) dose of radiation?

As you review the natural sources of radiation in this chapter and find out how radiation varies by geography and according to daily activities, there is no need to memorize specific dose values. But it is important to develop an informed perspective. Having such a perspective is the most practical way to ground your thinking when it comes to the subject of radiation.

RADIATION FROM NATURAL SOURCES

In this section, we describe the four major sources of ionization radiation to which people are constantly exposed:

- cosmic radiation;
- terrestrial radiation;
- internal radiation; and
- radon.

In technical jargon, the sum of radiation doses from these natural sources may be referred to as "background radiation" or "natural background radiation." Here, the word *background* simply means *baseline*.

Cosmic Radiation

Cosmic radiation, as the name implies, has extraterrestrial origins. It comes from the deep reaches of the universe (galactic particles) and from the Sun (solar particles). As these extraterrestrial particles strike the upper layers of the atmosphere, they produce a cascade of ionizations and shower the Earth with radiation. The atmosphere shields people from much of this cosmic radiation. At higher altitudes, where the atmosphere is thinner, cosmic radiation doses are higher than at sea level. For example, a person living in the Mile-High City (Denver, Colorado) receives approximately 0.47 mSV (47 mrem) of cosmic radiation annually, whereas a person living on the lower coasts of the United States receives 0.26 mSv (26 mrem) of cosmic radiation each year.

The average annual dose of cosmic radiation for the world population, allowing for variations in latitude and altitude, is approximately 0.4 mSv (40 mrem) per year (Table 2.1). The cosmic radiation levels in Mexico City, Mexico, and La Paz, Bolivia, are two and five times higher than this global average, respectively, and even higher in such countries as Tibet.

Did you know? A flight from Paris to San Francisco delivers a cosmic radiation dose of approximately 0.09 mSv (9 mrem) to each passenger and crew member. Airline crew members receive an average annual dose of 3 mSv (300 mrem) from their occupation, depending on routes flown and total flying time.*

Did you know? The method of carbon dating, used by archeologists and others to estimate the age of artifacts containing organic matter, is made possible by cosmic radiation. Carbon dating is based on measuring the amount of radioactive carbon the object contains. Radioactive carbon is produced initially in the Earth's atmosphere by collision and interaction of cosmic radiation with nitrogen. These radioactive carbon atoms ultimately end up as parts of trees, plants, oceans, and all living organisms, including us.†

---

* For an extended bibliography and a recent update on how the issue of occupational dose to airline crew members is addressed by scientific and regulatory organizations, see Barish, R. J. 2009. Health physics and aviation: Solar cycle 23 (1996–2008). *Health Physics* 96 (4): 456–464.

† The ratio of radioactive carbon and stable carbon in living organisms remains relatively constant; about 0.0000000001% of all carbon atoms are radioactive carbon. Once a plant, animal, or human dies, radioactive carbon in the body decays slowly with time and the ratio of radioactive to stable carbon atoms decreases even further because the organism is no longer alive to incorporate new radioactive carbon atoms. By measuring the ratio of radioactive to stable carbon atoms in the object, it is possible to estimate the object's age up to approximately 50,000 years.

Table 2.1 Worldwide Annual Radiation Doses from Natural Sources

Source	Average Dose (mSv/year)[a]	Typical Range (mSv/year)
Space	0.4	0.3–1.0
Earth	0.5	0.3–0.6
Human body	0.3	0.2–0.8
Radon	1.2	0.2–10
Total (rounded)	2.4	1–10

Source: Adapted from United Nations Scientific Committee on the Effects of Atomic Radiation (UNSCEAR). 2000. *Sources and effects of ionization radiation.* Report to the General Assembly, with scientific annexes. New York: United Nations.

[a] Multiply each number by 100 to convert the units to millirems per year.

Terrestrial Radiation

Terrestrial radiation originates from radioactive atoms that are naturally present in rocks and soil. Radioactive elements such as uranium, thorium, and potassium, which are present in the Earth's crust, have been around since the planet was formed and continue to irradiate people today as we stroll in neighborhood parks or rest at home. If atomic archeology were a field of study, researchers would need only to dig in their backyards or vegetable gardens to collect radioactive atoms that witnessed the Earth's formation more than four billion years ago.

Because soil composition varies from region to region, the dose a person receives from terrestrial radiation also varies by region. In the United States, the dose from terrestrial radiation, averaged by state, is 0.21 mSv (21 mrem) per year (Table 2.2). Orlando, Florida, has one of the lowest levels of terrestrial radiation in the United States. Residents of Orlando receive only 0.07 mSv (7 mrem) of radiation per year from the soil, which is three times less than the national average.* Another tourist destination, Las Vegas, Nevada, also has a low level of terrestrial radiation at 0.13 mSv (13 mrem) per year below the national average. Interestingly, Denver residents who already get a higher annual dose from cosmic radiation, receive a higher than average

* Bogen, K. T., and Goldin, A. S. 1981. Population exposures to external natural radiation background in the United States. U.S. Environmental Protection Agency, ORP/SEPD-80-12, Table A-2.

Table 2.2 Annual Radiation Doses to U.S. Population from Natural Sources

Source	Average Dose (mSv/year)[a]	Percentage of Annual Dose
Space	0.3	11
Earth	0.2	7
Human body	0.3	9
Radon	2.3	73
Total (rounded)	3.1	100

Source: Data from National Council on Radiation Protection and Measurements (NCRP). 2009. Ionizing radiation exposure of the population of the United States. NCRP report no. 160. Bethesda, MD.

[a] Multiply each number by 100 to convert the units to millirems per year.

radiation dose from soil and rocks as well—an average terrestrial radiation dose of 0.57 mSv (57 mrem) per year.

Similar variations in terrestrial radiation dose levels exist across the world. Some regions with unusually high levels of terrestrial radiation have monazite sand deposits with high levels of radioactive thorium—for example, the Guarapari area of Brazil, Yangiang in China, the southern states of Kerala and Tamil Nadu in India, and Egypt's Nile Delta. Some other high terrestrial radiation areas have volcanic soils, such as Mineas Gerais in Brazil, Niue Island in the Pacific, and parts of Italy. Some areas in central and southwestern France have granitic rocks and sands or uranium mineral deposits that emit higher radiation levels. Isolated areas in the Iranian cities of Ramsar and Mahallat have built-up radium deposits from hot springs that contribute to higher natural radiation levels. The average radiation levels could be 30 times the global average at these locations and even higher in isolated spots.*

Internal Radiation

The source of this type of natural radiation is inside the body. Because the food we eat, the water we drink, and the air we breathe contain naturally occurring radioactive materials, our body tissues contain several types of

* United Nations Scientific Committee on the Effects of Atomic Radiation (UNSCEAR). 2000. Sources and effects of ionization radiation. Annex B, Exposures from natural radiation sources, Table 11. New York: United Nations.

radioactive elements at all times. Most of the internal radiation dose comes from radioactive potassium. However, the body also contains radioactive elements such as uranium, radium, radioactive lead, and polonium. The radiation dose we receive from radioactive material inside our bodies is approximately 0.3 mSv (30 mrem) per year.

> *Did you know?* If you sleep next to someone, radioactivity contained in that person's body will irradiate you throughout the night. On average, you would receive an additional 0.01 mSv (1 mrem) of radiation per year from that person.
>
> *Did you know?* Out of every 10,000 atoms of potassium present in nature, one atom is naturally radioactive. Therefore, every substance that contains potassium is bound to contain radioactive potassium. For example, salt substitutes containing potassium chloride are rich in radioactive potassium. Many foods and vegetables contain varying amounts of radioactive potassium. Some examples are bananas, carrots, spinach, sweet potatoes, lima beans, peas, Brazil nuts, and coconuts.*

Radon

Radon gas is the largest natural source of radiation that the lungs receive and is believed to contribute to lung cancer.† Colorless, odorless radon gas is ubiquitous in the environment. Radon emanates from radium, a naturally radioactive material in soil, and is released into the air. Radon concentrations in outdoor air are relatively low because the gas disperses rapidly. However, if radon gas seeps into a house and builds up in an enclosed space, such as a poorly ventilated basement, radon concentrations may become much higher there and present a significant health risk when that air is breathed. If your basement is poorly ventilated and you and your family spend time there, it is prudent to have your house (especially the basement) tested for radon.

Radiation doses from radon vary significantly from region to region. Even within the same neighborhood, radon levels can vary greatly from one house to the next. On average, people in the United States receive a radon radiation dose estimated at 2.3 mSv (230 mrem) per year. The worldwide average is estimated at 1.2 mSv (120 mrem) per year.

For the U.S. population, the average annual dose from all sources of natural radiation is 3.1 mSv (310 mrem). Worldwide, this average has a typical range between 1 and 10 mSv (100–1,000 mrem).

---

* Klement, A. W., Jr., ed. 1982. Handbook of environmental radiation. Boca Raton, FL: CRC Press.

† Smoking cigarettes, the leading cause of lung cancer, significantly raises the risk posed by high concentrations of radon gas.

RADIATION FROM MAN-MADE SOURCES

Now that you have gained an informed perspective on naturally occurring background radiation, let us briefly compare those sources with man-made sources. The term "man-made radiation" refers to radiation that results from any human activity—accidental or intentional. This section reviews some of the typical sources of man-made radiation and places them in context with the natural sources just discussed.

Medical Sources

Medical procedures and devices that use ionizing radiation are the most important sources of man-made radiation. All other sources are a distant second. In the 1980s, it was estimated that Americans received an average annual radiation dose of 0.53 mSv (53 mrem) from diagnostic x-ray and nuclear medicine procedures. Two decades later, that estimate had increased nearly sixfold to 3 mSv (300 mrem) per year (Figure 2.1).*

In 2006, an estimated 395 million imaging procedures involving x-rays (excluding dental procedures) and nuclear medicine procedures were performed in the United States. Conventional radiography procedures (such as x-rays) were the most frequent examination, with 293 million procedures performed that year. However, these procedures contributed only 11% to the total medical radiation dose (Figure 2.2). The largest contributor to total medical dose was computed tomography (CT) scans (Figure 2.3).

Most people, however, do not receive x-rays or CT scans or have nuclear medicine procedures every year. Therefore, the "average population dose" does not reflect what any single individual is likely to receive. For a healthy young person, the medical radiation dose is likely to be zero for many years to come. But the sixfold increase in average population dose, in only two decades, indicates a clear trend. The U.S. population today receives as much radiation dose from medical procedures as it does from all other natural and man-made sources combined, including radon (Figure 2.4). Typical radiation doses for a number of common medical and dental procedures, as reported in the medical literature, are given in Table 2.3.

Even though the use of radiation in medicine is widespread and growing, there are significant disparities in available medical resources and the level of health care across countries. As a result, on a worldwide scale, medi-

* National Council on Radiation Protection and Measurements (NCRP). 2009. Ionizing radiation exposure of the population of the United States. NCRP report no. 160. Bethesda, MD.

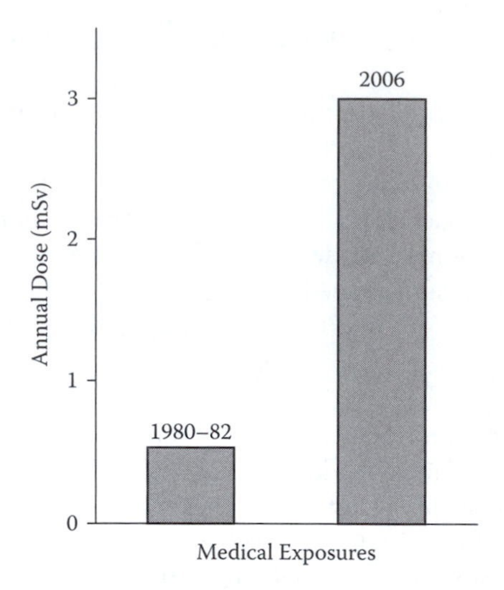

Figure 2.1
Annual medical radiation exposures for the U.S. population, averaged per person, increased nearly sixfold from the early 1980s to 2006. (From NCRP Report 160, 2009.)

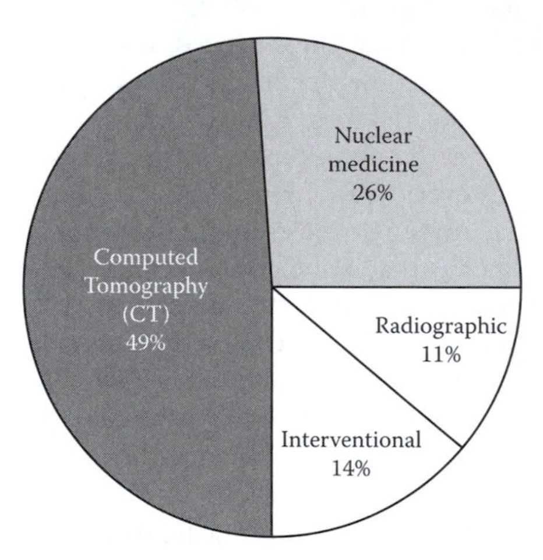

Figure 2.2
Percent contribution from various medical imaging and nuclear medicine procedures to the total medical radiation dose to the U.S. population. In 2006, an estimated 395 million such procedures were performed. (From National Council on Radiation Protection and Measurements. 2009. NCRP Report 160, Bethesda, MD.)

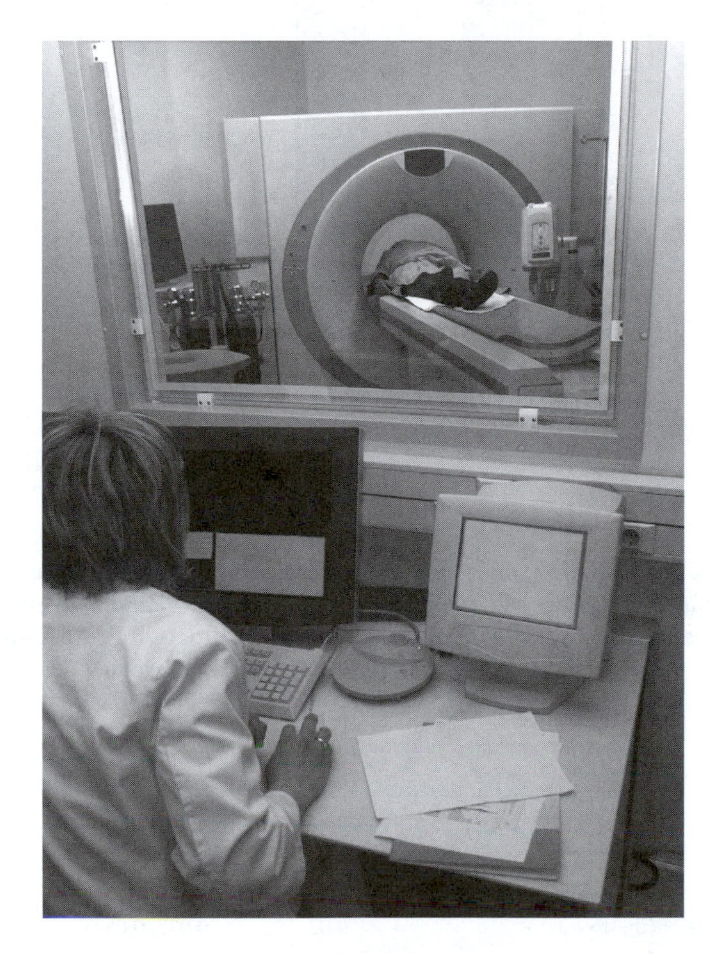

Figure 2.3
CT scans contributed nearly half of the total radiation dose from all medical procedures combined in 2006 (excluding radiation therapy). That year, 67 million CT scans were performed in the United States.

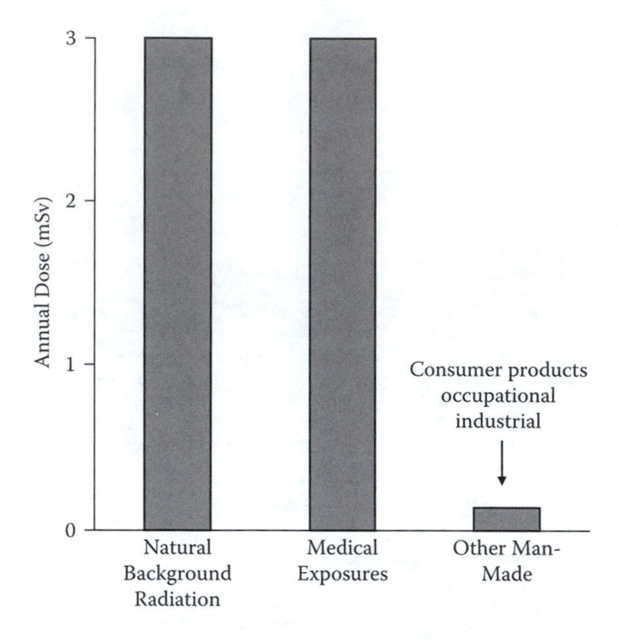

Figure 2.4

The U.S. population, on average, receives as much radiation dose from medical procedures as it does from all other natural and man-made sources combined. (From National Council on Radiation Protection and Measurements. 2009. NCRP Report 160, Bethesda, MD.)

Table 2.3 Typical Radiation Doses from Medical and Dental Procedures

Procedure	**Average Dose (mSv)**[a]	**Reported Range (mSv)**
Panoramic dental x-ray	0.01	0.007–0.09
Chest x-ray	0.02	0.007–0.05
Mammography	0.4	0.1–0.6
Barium enema	8	2–18
CT abdomen	8	3.5–25
Virtual colonoscopy	10	4–13
Coronary angiography	16	5–32

Source: Adapted from Mettler, F. A., W. Huda, and T. T. Yoshizumi. 2008. *Radiology* 248 (1): 254–263.

[a] Multiply each number by 100 to convert the units to millirems.

cal radiation contributes much lower doses in developing countries than in industrialized countries.

> *Terminology:* A nuclear medicine procedure is one that involves injection of radioactive materials inside the body, such as in a cardiac stress test. This procedure is different from a radiology procedure, which involves irradiating patients with x-rays. CT scans are sophisticated sectional x-ray exams that yield a three-dimensional image.
>
> *Did you know?* There are CT protocols specifically for children so that the same quality of diagnostic image is possible while delivering only "kid-size" radiation doses. Not every imaging facility, however, uses these pediatric protocols.
>
> *Did you know?* Diagnostic imaging procedures such as ultrasound and magnetic resonance imaging (MRI) do not use ionizing radiation.
>
> *Did you know?* It would not be unusual to receive a measurable radiation dose from someone sitting next to you on a subway or bus if that person has had an outpatient nuclear medicine procedure in the past few days (i.e., the patient was released from a hospital after being injected with radioactive material for examination or treatment).

Other Man-Made Sources

Decades ago, the atmospheric testing of nuclear weapons resulted in unrestricted release of a large quantity of radioactive materials into the atmosphere; these were widely dispersed and deposited everywhere on the Earth's surface. Fortunately, their radioactivity levels dropped significantly before the materials were deposited on the surface and have continued to decrease in the 40 years since atmospheric testing stopped. As a result, the residual nuclear fallout contributes less than 0.005 mSv (0.5 mrem) to an average annual radiation dose today.*

Many consumer products and materials that people routinely encounter contain radioactive materials. The proper functioning of smoke detectors depends on radioactive americium, and self-luminous exit signs contain tritium. Even ceramic tiles, bricks, and granite naturally contain radioactive elements such as uranium and thorium. As a result, workers in New York City's Grand Central Station and the U.S. Capitol Building in Washington, D.C., for example, receive higher than average annual doses of radiation because of natural uranium in those structures' granite walls. Leaving these examples aside, the average annual dose of ionizing radiation received from these and other miscellaneous sources—in addition to radioactivity released

* This global average estimate is from the UNSCEAR 2000 report. The average dose from nuclear fallout was 0.15 mSv (15 mrem) in 1963—30 times higher than the estimate in 2000—because the radioactivity levels have continued to decrease.

by nuclear fuel processing, nuclear power plants, and coal-burning electric power plants—amounts to <0.2 mSv (<20 mrem).

Radiation dose values reported here do not include accidental releases or accidental exposures, which are discussed in Chapter 4. Furthermore, these radiation dose estimates are averages. For example, smokers receive significantly higher radiation doses to their lungs. Tobacco plants trap naturally occurring airborne radioactivity (radioactive lead and polonium) on the surface of their leaves. When tobacco leaves are smoked, these radioactive elements are volatized and inhaled into the lungs. Radiation doses to the bronchial epithelium (the inner lining of the lung) from tobacco leaves' natural radioactivity can be quite large and pose additional, serious health risks for smokers.

WHAT SHOULD YOU DO?

Radioactivity is an inseparable part of nature and exposure to radiation is a part of life. Knowledge of natural background radiation should not cause people to relocate to a different part of the country, quit jobs, cancel vacation plans to a mountain or ski resort, refuse to fly on commercial jets, or stop taking strolls in a neighborhood park. Similarly, developing an understanding of the sources of man-made radiation should not lead someone to forego needed medical procedures, remove invaluable smoke detectors from the home, or avoid buildings that contain ceramic tiles, bricks, or granite. These sources of radiation, with all their variations from place to place, are accepted as part of the natural and man-made environment and are not regarded as health threats.

This does not mean, however, that every small dose of radiation should be accepted without question. A small additional dose of radiation may be insignificant from an immediate health perspective. But if the dose of radiation does not provide society with any medical, occupational, or recreational benefits, it is not a justifiable dose of radiation and should be avoided, if possible, using reasonable means.

You can, and should, take some very important steps to minimize exposure to radiation that does not have offsetting benefits:

- No one should smoke. Tobacco use is the leading cause of premature deaths every year in the U.S.* To make things worse, tobacco

* Mokdad, A. H., J. S. Marks, D. F. Stroup, and J. L. Gerberding. 2004. Actual causes of death in the United States, 2000. Journal of the American Medical Association 291 (10): 1238–1245, with a correction by these authors in Journal of the American Medical Association 293 (3): 293–294, 2005.

leaves and smoke contain radioactivity and can deliver relatively high doses to specific regions of the lungs. Smoking also increases vulnerability to radon gas.

- You can test your home for radon. If you use an underground basement as living space, it is prudent to test the basement for radon. Radon testing is inexpensive; if radon concentrations in the house are high, in most cases, effectively correcting the problem is also relatively inexpensive.
- You should keep track of medical exposures—for example, how many CT scans you have had. This point is especially important if you see different care providers for the same conditions. Your medical care providers should be aware of your cumulative exposures and, if indicated and feasible, use alternative methods of diagnosis, as long as your care is not compromised. Also, if your child needs a diagnostic CT scan, ask your doctor or the radiology service professional whether he or she uses pediatric protocols for imaging. Pediatric protocols typically use a much smaller radiation dose (by a factor of two or more) for the same quality image.

Resources

The following two reports provide detailed information about natural and man-made sources of radiation in various countries as well as the United States:

United Nations Scientific Committee on the Effects of Atomic Radiation (UNSCEAR). 2000. Sources and effects of ionization radiation. Report to the General Assembly, with scientific annexes. New York: United Nations, New York (available from www.unscear.org/unscear/en/publications/2000_1.html).

National Council on Radiation Protection and Measurements (NCRP). 2009. Ionizing radiation exposure of the population of the United States. NCRP report no. 160. Bethesda, MD. (This is an update to NCRP report 93—Ionizing radiation exposure of the population of the United States, 1987.)

The following is a classic textbook on the subject of environmental radioactivity. Natural radioactivity is described in its Chapter 6: Eisenbud, M., and T. Gesell. 1997. *Environmental radioactivity from natural, industrial, and military sources.* San Diego: Academic Press.

This following book provides a fascinating illustrated history of radioactivity from natural sources and commercial products: Frame, P., and

W. Kolb. 2002. *Living with radiation: The first hundred years,* 3rd ed. Maryland: Syntec, Inc. (The book is self-published by the authors.)

The World Health Organization (WHO) offers valuable information on radon and health effects of exposure to radon (www.who.int/ionizing_radiation/env/radon/en/index.html).

The U.S. Environmental Protection Agency Web site (www.epa.gov/radon) offers a number of resources for the public regarding radon, including the citizens' guide. U.S. Environmental Protection Agency. 2009. *A citizen's guide to radon.* EPA 402/K-09/001. Available online (www.epa.gov/radon/pubs/index.html).

In the United States, you can find the contact information for your state's radon coordinator at www.crcpd.org/Radon.asp

For an in-depth scientific review of available literature on health effects of radon, see this report by the National Academy of Sciences: National Academy of Sciences. 1999. *Health effects of exposure to radon.* BEIR VI report. Washington, D.C.: National Academies Press.

The U.S. Environmental Protection Agency provides a simple online calculator where you can find your estimated yearly dose from natural sources of radiation (www.epa.gov/rpdweb00/understand/calculate.html).

To obtain an estimate of cosmic radiation dose from a specific flight itinerary, use the online calculator at the Federal Aviation Administration (FAA), Office of Aerospace Medicine, Civil Aerospace Medical Institute Web site (http://jag.cami.jccbi.gov/cariprofile.asp).

The computer code CARI-6 can also be downloaded from the FAA Web site (www.faa.gov).

Information about a variety of radiation imaging and therapy procedures is available at www.radiologyinfo.org

Regarding the subject of radiation safety in pediatric imaging, the Image Gently Campaign provides information for patients and health care providers. This information is available from the Society of Pediatric Radiology Web site (www.pedrad.org).

The Society for Pediatric Radiology and the National Cancer Institute jointly produced *Radiation risks and pediatric computed tomography (CT): a guide for health care providers* (available from www.cancer.gov).

For referring physicians, radiologists, and other health care providers, the American College of Radiology Appropriateness Criteria® provides guidelines for making the most appropriate imaging or treatment decisions (available from www.acr.org/ac).

Recommendations of an American College of Radiology blue ribbon panel to address the issue of radiation dose in medicine were published in this article: Amis, S. E., P. F. Butler, K. E. Applegate, Birnbaum, S. B., Brateman, L. F., Hevezi, J. M., Mettler, F. A., et al. 2007. American College of radiology white paper on radiation dose in medicine. *Journal of the American College of Radiologists* 4:272–284.

The European Commission published *Referral guidelines for imaging* as Radiation Protection Series 118 in 2000. These referral guidelines, which are not binding, are available online from the European Commission Web site (http://ec.europa.eu/energy/nuclear/radioprotection/publication/doc/118_en.pdf).

3

RADIATION 101

OVERVIEW

RADIATION 101 IS NOT THE MOST titillating topic. Do you really need to hear about the oscillating electric and magnetic fields of electromagnetic radiation or know that alpha particles are made of two protons and two neutrons, equivalent to the nucleus of a helium atom? Do you really need to know how the photoelectric effect works when photons strike the atom, and how Albert Einstein's explanation of that phenomenon contributed to the quantum revolution in physics and won him the Nobel Prize? Not in this book! But you do need to know how alpha particles or gamma-ray photons can affect you differently, how you can tell them apart, how far they can reach, and how you can stop them or protect against them. You also need to know how people can be exposed to radiation but not get contaminated.

The ultimate motivation for most readers of this book is to learn how to protect against radiation. If you understand the physical nature of what you face and some basic fundamentals, you can provide yourself with better and smarter protection. To that end, this chapter focuses on the information you need to know and provides examples to help explain important concepts. We avoid the detailed physical descriptions and mathematical formulations that tend to excite radiation enthusiasts. Readers who wish to find such information can use a number of resources offered at the end of the chapter.

We will explore what is meant by "radiation" or "radioactive decay," and we will answer these questions: What types of radioactivity exist and how can they can affect you? What is isotope identification and why is it important? What is half-life? How do you use this knowledge to protect yourself? What is the difference between radioactive contamination and "exposure" or irradiation?

Familiarity with these fundamentals can offer a better grasp and comprehension of radiation-related information you may hear or read elsewhere.

Some readers may be inclined to skip this chapter. If you do, come back to it as you feel the need. Unfamiliarity with these basic concepts will limit what you gain from the rest of the book.

WHAT IS RADIOACTIVITY?

Some atoms found in nature, especially the heavy ones such as uranium, are unstable. An unstable atom has extra energy bottled inside and must get rid of that energy to reach a stable state. As an example, imagine yourself after drinking a can of carbonated beverage when carbon dioxide has built up in your stomach and is looking for a way out. When the time comes, an unstable atom spontaneously ejects the extra energy from its nucleus and reaches a more stable form. This spontaneous process of releasing energy is called radioactivity, and the unstable atom emitting it is said to be radioactive. This excess energy is ejected in the form of radiation or, more accurately speaking, in the form *ionizing* radiation. It is the ionizing nature of this radiation that sets it apart from all other types of radiation.

> *Did you know?* Radioactivity is not just for large heavy elements. Smaller atoms, readily found in nature, come in unstable forms too. Some atoms can exist in multiple forms called isotopes; some isotopes are stable and some are unstable. Potassium and carbon are good examples of atoms where small fractions of total potassium and carbon, abundantly found in nature and in the body, are radioactive and emit ionizing radiation to achieve a more stable form.

WHAT IS RADIATION?

The word "radiation" is a broad term that, technically speaking, includes radio waves, microwave radiation, radar, infrared radiation (or heat), visible light, and ultraviolet light (UV radiation), as well as x-rays and gamma rays. When most people talk about radiation, obviously they are not talking about radio waves. The type of radiation that creates the most anxiety is the type you may associate with atomic bombs, nuclear power plants, x-ray machines, or space travel. In films and fiction, it is the same type of radiation that leaked from the warp drive and killed Mr. Spock or affected the fictional physicist, Dr. Bruce Banner, thus creating the Incredible Hulk.

What sets this type of radiation apart from all others is its ability literally to knock electrons out of atoms and make ions. This effect is called ionization and can occur in any material—lead, steel, water, plastic, or human tissue.

In fact, these types of ionizations are occurring in your body right now as you read these lines (see discussion of internal radiation in Chapter 2). As we said earlier, this type of radiation is more accurately referred to as ionizing radiation; however, for simplicity, we just call it radiation.

The type of radiation that concerns us in this book is the energy ejected from an unstable radioactive element. This radiation comes in one of two forms: (1) packets of energy that travel through space at the speed of light in the form of waves, or (2) small subatomic particles that are ejected at high speed.

Often, when a radioactive atom ejects its excess energy, it does so by emitting some of both forms of radiation: waves and particles. The wave radiation includes gamma rays and x-rays.* The particle radiation includes alpha particles, beta particles, and neutrons. Each of these has different properties and protection against them requires different strategies. We briefly describe each form and its properties in this section.

Gamma Rays and X-rays

Gamma rays and x-rays are electromagnetic waves that come in discrete packets of energy called photons. In the physical world, the shape and nature of x-rays and gamma rays are similar to other, more familiar forms of electromagnetic waves such as radio waves, microwaves, radar, visible light, and infrared. Although this statement is a scientific fact, it can be confusing and a bit misleading. Photons from x-rays or gamma rays interact with matter differently because they pack much more energy than other forms of electromagnetic radiation—1,000 times more than UV radiation, 1,000,000 times more than infrared radiation, 1,000,000,000 times more than microwaves, and 1,000,000,000,000 times more than radio waves.

Because they can penetrate and even pass through objects as they interact with them, x-rays and gamma rays are considered penetrating radiation. That is why they are used in medicine to image the inside of the body. They go through the soft tissues with ease and strike the photographic film or plate placed behind or underneath the body. But the bones in the body absorb the x-rays and that creates a shadow (image) on the film.

If unobstructed, x-rays and gamma rays can reach the length of a football field (10s of meters) in air before they lose their energy and dissipate. The best material to absorb and stop x-rays or gamma rays is lead—hence the lead apron used in medical offices. Even in a high-level gamma radiation field, a few inches of lead can absorb the radiation and shield from it completely.

* The x-rays encountered the most are produced by machines, rather than emitted by radioactive atoms. But because x-ray properties are similar to those of gamma rays, we include them in this discussion.

Lead containers often are used to contain radioactive sources during shipment or storage and such containers work quite well. Firefighter turnout gear or any other type of garment cannot protect against x-rays and gamma rays.

What Is the Difference between X-rays and Gamma Rays? Physically speaking, x-rays and gamma rays do not differ, except that gamma rays usually have higher energy and are more penetrating. The reason for differentiating between the two is because of where the photons originate. Gamma rays originate from the nucleus of an atom when the nucleus changes to a more stable energy state. On the other hand, x-rays originate from electron transactions, rather than from the nucleus. Once they are both produced and leave the atom, there is no physical difference between x-rays and gamma rays. The x-rays that are encountered routinely, such as dental or chest x-rays, are produced by machines. Hospital computed tomography (CT) machines also generate x-rays to produce images. Once the machine is turned off, no x-rays are produced.

Gamma rays are produced naturally by radioactive atoms and can have higher energy than x-rays. Many radiation therapy machines contain a large radioactive source that emits gamma rays all the time, day and night. When the machine is turned off, the radioactive source is automatically rotated and kept inside a thick lead housing. The source still radiates gamma rays, but the gamma rays are all absorbed by the lead shield and do not exit the therapy machine. When the machine is turned on, the source is rotated and brought outside the lead shield. The gamma rays are allowed to come out unobstructed to irradiate the patient.

Some of the gamma rays striking the patient bounce back, or ricochet, in different directions and are called "scatter" radiation. Scatter radiation always has much lower intensity than the primary beam of radiation. Nevertheless, the nurse or the technologist helping the patient leaves the room or stands behind a shield when operating these or similar machines so that he or she is not hit by the scatter radiation.

Particulate Radiation

This type of radiation comes in the form of highly energetic, fast-moving subatomic particles. How they behave depends on their size, the energy they carry, and whether they carry a negative or positive electrical charge. Three major types of particulate radiation are of concern: alpha and beta particles, and neutrons.

Alpha Particles In the subatomic scale, alpha particles are bulky, have a positive charge, and travel slowly. As a result, alpha particles bump into and interact with any matter they encounter, give up their energy, and disappear rapidly. If unobstructed, alpha particles can travel only a few centimeters in air (2–3 inches). A single sheet of paper placed in the track of an alpha particle will stop it completely. If alpha particles strike human tissue, they can penetrate only a short distance of 0.04 millimeters (mm) (1/1000 of an inch).

As long as alpha particles remain outside the body, even directly placed on the skin, they present no hazard at all because they cannot penetrate the body. The dead outer layer of skin covering the body is more than adequate to block all alpha particles. However, if alpha particles are emitted inside the body, they can be a serious health hazard. This can occur if the radioactive material emitting the alpha particles is aerosolized and inhaled, somehow ingested, or absorbed through an open wound. In this case, the individual is said to be internally contaminated, a subject we describe later in this chapter.

> *Did you know?* A radioactive atom ejects an alpha particle at initial speeds approaching 15,000 kilometers (km) per second (more than 9,000 miles per second). Yet the particle can be blocked completely by a thin layer of material. As fast as that initial speed may seem, it is comparatively slow in the subatomic world—much slower than the speed of light.

Beta Particles Beta particles are energetic electrons ejected from the nucleus of a radioactive atom. They carry a negative charge and are much smaller than alpha particles—more than 7,000 times smaller in fact. Because they are so small, beta particles are a bit more penetrating than alpha particles. Still, beta particles can be easily stopped by a thin sheet of aluminum a few millimeters (1/10 of an inch) thick, or a piece of lucite (Plexiglas) that is about 1.5 centimeters (cm) (or half an inch) thick. Even if no shielding is available, beta particles can travel only a short distance in air, losing their energy and intensity every step of the way. The distance they can travel depends on their energy, but it is typically only a few meters (m) (several feet) if unobstructed.

When beta particles encounter human skin, they can penetrate to a distance of several millimeters (between 1/4 and 1/2 inch). This means that beta particles can irradiate the live skin cells underneath and, if the amount of beta particle radiation is high enough, can cause serious skin injury, sometimes called beta burns. The good news is that clothing offers protection against beta particle radiation.

Just as with alpha particles, beta particles can cause the most severe health effects if they are emitted inside the body. Again, this can occur if the radioactive material emitting the beta particles has been inhaled, ingested, or absorbed through a wound.

Neutrons Neutrons are subatomic particles about 2,000 times larger than electrons. However, because they carry no electric charge, they can pass

Table 3.1 Types of Radiation That May Be Present in a Contaminated Area

Radiation Type	Penetrating Ability	Hazard
Gamma rays	High	Similar to x-rays. Firefighter turnout gear offers no protection. A hazard both outside and inside the body.[a]
Alpha particles	None	Cannot penetrate skin. Significant hazard only if contamination is inside the body.
Beta particles	Low	Can cause serious skin injury, but cannot penetrate deeper into the body. Clothing offers some protection for the skin. Hazardous if contamination is inside the body.

[a] Contamination occurs inside the body if radioactive material is aerosolized and inhaled, somehow ingested, or absorbed through an open wound.

through a lot of material before interacting and losing their energy. Because of the unique way neutrons interact with matter, a high-density material such as lead does not necessarily make a better shield. A material rich in hydrogen is the best shield for neutrons. Concrete blocks, several feet thick, will stop neutrons. In fact, ordinary water, because it is rich in hydrogen, is also a good shield against neutrons. These shielding considerations are relevant in design and operation of a nuclear facility, but not for us.

Following a radiation incident, we do not have to worry about shielding against neutrons because neutrons are emitted only at the instant a nuclear accident occurs or a nuclear bomb detonates. When it is time to take protective actions, neutrons are no longer in that environment. The types of radiation we need to be concerned about and protect against are summarized in Table 3.1.

> *Did you know?* Neutron radiation does not randomly occur in nature. One element that emits neutrons is an isotope of a heavy element called californium. This element is not found naturally and must be synthesized. Californium does have some limited industrial and medical uses.
>
> *Did you know?* Trillions of subatomic particles called neutrinos* pass through your body every second without touching you. Most of these neutrinos originate from nuclear reactions within the Sun and shower the Earth at speeds near the speed of light, passing right through you and the ground you are standing on, emerging from the other side of the planet unscathed and without interacting with anything.

* Not to be mistaken with neutrons, these elementary particles have nearly zero mass.

Nomenclature

The radioactive form of any element is called a radioisotope or radionuclide. Either term can be used. Alpha and beta particles are often written using Greek alphabetical characters, as in α-particles and β-particles. Gamma rays are also written using the Greek alphabetical character, as in γ-rays. Neutrons may be designated by a lowercase, italicized *n*.

Radionuclides are written in a variety of ways using the chemical name or symbol from the periodic table and a number to designate the specific isotope. It is common to put the atomic mass number in superscript to the left of the symbol or just spell out the name of the element and hyphenate the atomic mass number. For example, the radioactive element uranium can be represented by uranium-238, ^{238}U, or simply U-238. The designated number next to the radionuclide does not mean that over 200 isotopes of uranium exist. It just refers to the atomic mass of that particular radionuclide; the larger this number is, the bigger the atom is. Uranium does have several radioactive isotopes, notably U-234, U-235, and U-238. They are all radioactive and emit alpha particles and gamma rays. There are no stable forms of uranium.

> *More examples:* The most abundant form of carbon is the stable C-12 atom. The naturally occurring C-14 atom is a radioactive isotope of carbon and emits low-energy beta particles. The most abundant form of potassium is the naturally occurring stable K-39 atom. The naturally-occurring K-40 atom is a radioactive isotope of potassium that emits beta particles as well as weak gamma rays. The heaviest stable metal found in nature is lead in the form of Pb-206 and Pb-207. The radioactive element Pb-210, which is also readily found in nature, emits beta particles.

You do not have to commit any of this information to memory, of course. You just need to know that the information is available from reference tables or you can ask a radiation professional. Table 3.2 lists a few examples of radionuclides, the primary type of radiation they emit, and their half-lives, which we describe later in this chapter.

Why Is Radionuclide (or Isotope) Identification Important?

In a radiation accident or facing a radiation emergency, one of the first questions the responders should ask is which specific radionuclide is involved. The characteristics of the radionuclide will be a determining factor in how the responders protect against it, what instruments should be used to detect the radiation, and what environmental protection criteria will apply. In short, the entire response and recovery operation is influenced by the type of radionuclide involved as well as the amount of radioactivity.

Table 3.2 Examples of Radionuclides

Radionuclide	Primary Radiation	Half-Life
Americium-241 (Am-241)	Alpha	430 years
Cesium-137 (Cs-137)	Gamma, beta	30 years
Cobalt-60 (Co-60)	Gamma, beta	5.3 years
Iridium-192 (Ir-192)	Gamma, beta	74 days
Iodine-131 (I-131)[a]	Gamma, beta	8 days
Plutonium-239 (Pu-239)	Alpha	24,000 years
Radium-226 (Ra-226)	Alpha, gamma	1,600 years
Strontium-90 (Sr-90)	Beta	29 years
Uranium-235 (U-235)	Alpha, gamma	704 million years
Uranium-238 (U-238)	Alpha	4.5 billion years
Technetium-99m (Tc-99m)[b]	Gamma	6 hours

[a] I-131 is used in medicine for treatment of thyroid cancer.

[b] Tc-99m is widely used in medical diagnostic procedures, partly because of its short half-life.

RADIOACTIVE DECAY

Earlier, we described the process when an unstable atom spontaneously releases some of its energy to become stable. That process is referred to as radioactive decay or disintegration. The words "decay" and "disintegration" are not the best words to describe this phenomenon because the material does not go away or fall apart. It just changes shape to a more stable form. For example, when a carbon-14 atom "decays" by ejecting a beta particle, it changes form to become a stable nitrogen atom. When polonium-210 decays by ejecting an alpha particle, it changes shape to a stable atom of lead. Sometimes the words "transition" or "transformation" are used to describe the process, but most of the scientific and technical community still use radioactive "decay."

Parents and Daughters

Some radioactive atoms have to go through a series of transformations (decays) to reach a stable form ultimately. One radionuclide emits radiation and changes into a lower energy atom, but still in an unstable state. The newly formed radionuclide emits radiation again to reach a yet more stable state. The first radionuclide is said to be the "parent" radionuclide. The second is called the "daughter" product or progeny. This process may be repeated many times until finally a stable atom is formed. In other words, a daughter product serves as parent radionuclide to another daughter product and so on. This series of tandem transformations is called a "decay series" or a "decay chain." Uranium and thorium are good examples. Much of the radioactive elements found naturally on Earth are members of the natural uranium and natural thorium decay series. The most notable example is radon gas, Rn-222.

How "Hot" Is It?

Most people have heard the expression "hot" to describe radioactive materials, but very few people know what it actually means. It is true that the process of radioactive decay produces some heat, but the expression "hot" has nothing to do with temperature. It is a way of expressing how radioactive the material is. This is expressed scientifically by how many radioactive decay processes are happening every second. (See the section on radiation units for more information.)

> *Did you know?* A lethal radiation dose of 100 Gy (10,000 rads) to the whole body deposits enough ionization energy in the body to raise its temperature by only 0.024°C.

HALF-LIFE

The half-life of a radioisotope is the amount of time it takes for half of its atoms to decay spontaneously. Let us illustrate this with an example: Start with a certain number of radon-222 atoms. The half-life of radon-222 is approximately 4 days. That means that sometime during the next 4 days, 50% of these atoms will give off their radiation and decay. In the following 4 days, half of the remaining atoms will also decay. That means that in 8 days (two half-lives), only 25% of the original radon atoms remain. After 28 days (seven half-lives), less than 1% of radon atoms will remain and the rest have decayed. The concept of half-life is illustrated in Figure 3.1. The half-life is very specific to each radioisotope and it can range from a fraction of a second to a few hours, days, decades, or even billions of years.

> *Rule of thumb:* No matter what the radionuclide is, after seven half-lives, the initial amount of radioactive material decreases to less than 1%.

Consider a certain number of uranium-238 atoms. The half-life of uranium-238 is 4.5 billion years. This means that you would have to wait that long—the approximate age of planet Earth—for 50% of the uranium atoms

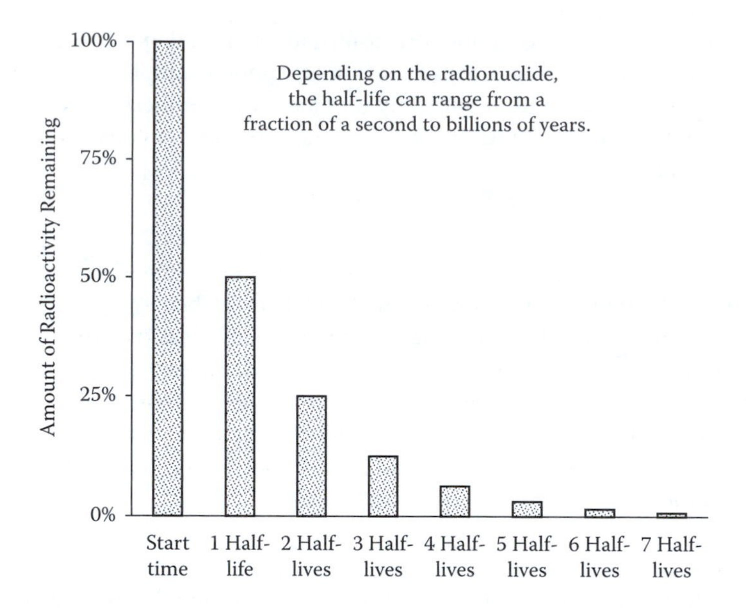

Figure 3.1

The concept of half-life. For all radioactive materials, the amount of radioactivity decreases with passage of time and drops by one-half after each successive half-life. After seven half-lives, the amount of radioactivity drops to less than 1% of the original amount.

to decay. Thus, uranium-238 decays at a much slower rate than radon in the previous example. Nearly all radioactive materials used in nuclear medicine have short half-lives measured in hours or days. Table 3.2 lists a sample of common radionuclides with commercial, industrial, military, and medical applications.

Biological Half-Life

In situations when radioactive contamination is inside the body, the term "biological half-life" refers to the time it takes for the body to excrete half of that contamination, mainly through urine and feces. For example, cesium-137 has a relatively long physical half-life of 30 years. Therefore, if the body does not excrete the cesium, the contamination will remain for decades. However, the biological half-life for cesium is about 110 days for adults (fewer days for children), so cesium contamination will be excreted from the body within several months. Depending on the amount of initial contamination, the harm done over that time period can be significant.

EXPOSURE (IRRADIATION) AND CONTAMINATION

The difference between radiation exposure and radioactive contamination is important, and it is one of the most challenging concepts to grasp because it takes a bit of "unlearning" to comprehend. Being exposed to a chemical or biological material means coming in contact with that material. If you are exposed to mercury, this means you came in contact with it on you or in you. If you are exposed to the flu virus—again, this means you came in contact with it. This is not the way the word "exposure" is used when radiation is discussed.

To appreciate the difference, you must mentally separate the source of radiation from the radiation itself. The source of radiation is the radioactive material—the gas, liquid, or solid object that contains the individual radioactive atoms. The radiation is the energy that the radioactive source is emitting in the form of rays and particles. The radiation being emitted can interact with you and do harm to your cells, even if you do not come in contact with the radioactive source itself.

In the vacuum of space, the radiation rays and particles can go far and travel a long time. But, on Earth, once they are emitted, all the rays and all the particles have an extremely short life span before they give up their energy and disappear—almost instantly. It is important to recognize that radioactive atoms do not disappear or go anywhere, but rather that the radiation they emit disappears in a flash after it leaves the atom.

If you are unfortunate enough to be in a position where these rays and particles can hit you in the instant that they are emitted, they have an opportunity to lose their energy in your body and potentially damage your cells. At that time, no remnants of the rays and particles are in or on your body: the rays and particles have vanished. Only the cellular damage they may have caused is left, and the culprit annihilated. This interaction can happen even if you do not make any contact with the source of radiation. In this sense, you were "exposed" to radiation—the rays and particles—but you did not come in contact and did not get "contaminated" with the radioactive material itself.

For example, suppose you accidentally sit next to a gamma-emitting radioactive source of cesium-137. Let us say that a glass window is between you and the source. The cesium atoms are behind the window and stay there, but the gamma rays they shoot will easily go through the glass and enter your body. As long as you are sitting against that window, the gamma rays keep coming and keep hitting you without your sensing them. Some of them go right through your body and hit the opposite wall in the room. Some of the rays interact with your cells and lose their energy inside your body, damaging your cells. But, without exception, every gamma ray coming from behind the window will disappear forever in a fraction of a second.

Note that, during all of this, you are not coming in contact with any cesium-137 atoms. If you get up and leave the area, you are certainly not carrying any cesium away with you. No rays or particles are hanging around your body or your clothes. You are not contaminated.

However, you were definitely exposed to the gamma radiation, and depending on how long you were sitting there and how much cesium was behind the window shooting at you, you could have a significant amount of damage in your cells. The damage could even be lethal, but the important thing here is that you were not contaminated with any radioactive material and will pose no danger whatsoever to people around you, including those who will provide you with medical care.

Let us use a more routine clinical example to illustrate the same point. When patients are irradiated with x-rays during a CT scan, they are exposed to the radiation that the machine produces. Once the machine is turned off and patients walk away, they do not carry any radioactivity on them or inside them. When patients go home after the CT scan, they do not carry any radiation or radioactivity home with them.

This act of irradiation is often referred to as "exposure" in the radiation world. This is an unfortunate choice of words, and until radiation scientists stop using this terminology, you have no choice but to learn it. Therefore, if you are exposed to radiation—whether it is rays or particles—it means you are being irradiated with rays and particles that dissipate in an instant. It does not mean you are contaminated with the source of that radiation.

Fortunately, the word "contamination" in the radiation world has the same familiar meaning that it has in everyday usage. Radioactive contamination means coming in contact with the radioactive material. Let us use the cesium-137 source again. Suppose this radioactive material is unshielded and does not have a warning label on it. Radioactive material can be a gas, liquid, or solid. In the case of cesium-137, it is solid and usually comes in the form of powder. Suppose you unwittingly pick up some of that cesium powder to examine it. At that point, you have come in contact with the source (the cesium), and your hands are contaminated with it. You have it on your skin. This is called "external contamination" because the source of radioactive material is still outside your body.

Now suppose you smell the powder, not knowing what it is. By smelling it, you might accidentally inhale a small amount of the cesium and contaminate your nostrils and your lungs with it. This is called "internal contamination" because the source of radioactive material is now inside your body.

Continuing with the example, you may lose interest in the source and walk away. Let us suppose that you have poor hygiene habits and eat a sandwich without washing your contaminated hands. This means that you will ingest some cesium-137 as well, adding to your internal contamination. The cesium atoms in your lungs and your stomach make their way to your blood and then are distributed throughout your muscles and fat cells. All the while, no matter where the cesium atoms are sitting, they are shooting off gamma rays and beta particles. They are doing so when cesium atoms are on your skin, in your lungs, and in your muscles. You are being irradiated from outside your body (cesium on your skin) and from inside your body. Therefore, when you are "contaminated" with radioactive atoms, you are also being "exposed" to the radiation that they emit.

An accident similar to the hypothetical example just described occurred in a Brazilian community in 1986. In that tragedy, several people were both exposed to and contaminated with cesium-137. The most severely affected was a 6-year-old girl who ate with contaminated hands. This accident is described in Chapter 4.

> *An example from the farm:* A common way of teaching this concept is to picture a person strolling in a pasture. If the person smells cow manure, he is being "exposed" to it. If he walks away from that area, he does not carry any cow manure with him. But if he happens to step in it, he becomes "contaminated" with cow manure and when he walks away, he will carry some of it with him. When contaminated with cow manure, he can also smell it, meaning he is exposed to it at the same time he is contaminated. This is external contamination because manure is on his boots, outside the body. This example is not a suitable one for illustrating internal contamination, but you get the picture.

Difference between Biological Contamination and Radioactive Contamination

When a person is contaminated with an infectious disease such as influenza, cholera, or plague, the disease will spread through contact among humans, and the spread may cause an epidemic. Radioactive contamination does no such thing. Imagine that your hands are covered with dust. As you shake hands with people, you transfer dust to them. The amount of dust you transfer decreases each time you shake hands with another person. Even though you spread dust to other people, the amount of dust never increases. You essentially dilute the amount of dust on your hands. Consequently, the 10th person shaking hands with you may not even realize that you had dusty hands because so little dust is left.

Radioactive contamination works in much the same way. Even if initial levels of contamination are high and several people develop radiation sickness, the sickness is not communicable and never turns into an epidemic.

Difference between Chemical Contamination and Radioactive Contamination

When a person comes in contact with a nerve agent, such as sarin gas, symptoms can appear within seconds and the situation is life threatening. Coming in contact with cyanide can also be immediately life threatening. Contamination with radioactive materials, however, rarely presents an immediate danger. Radioactive debris can be washed off or clothes removed. Most cases of radiation contamination involve low levels of radiation, except for nuclear fallout when the dust falling to the ground contains high levels of radioactivity, and the material should be washed off as quickly as possible. (See Chapter 9 for more discussion on this topic.)

Responders who train to protect against dangerous chemical and biological agents tend to apply the same safety principles to radioactive contamination. In many cases, this overprotection is not necessary and may even prove counterproductive.

RADIATION DETECTION

Radiation cannot be seen, heard, felt, tasted, or smelled. None of the senses can detect radiation, even though it has been around since Earth's beginning. But, with the right equipment, radiation can be easily detected. A variety of radiation instruments has been designed to detect gamma rays, alpha particles, and beta particles. Specialized instruments are used to detect neutrons as well. Many of these instruments are portable and can be used in the field to detect the presence of radiation in the environment or on a person.

The most versatile equipment is the familiar Geiger–Muller (GM) radiation detector. These instruments are relatively easy to operate, but they should be used by a trained and knowledgeable individual. If a layperson uses this equipment without proper training, he or she can potentially misinterpret the reading, reach an incorrect conclusion, and make an inappropriate decision.

Using these common portable field instruments, it is possible to say whether alpha, beta, or gamma radiation or any combination of the three is present—but not which radionuclide it is. In the case of gamma emitters, specialized portable instrumentation can identify the specific radionuclide in the field without the need for laboratory analysis. These instruments are called gamma spectrometers. In a laboratory, it is possible to detect minute quantities of radioactivity in soil, water, air, urine, or food samples.

Most people are familiar with units of measure such as parts per million (ppm) and parts per billion (ppb), which are used to describe chemical contaminants. For some radionuclides, laboratory equipment is able to see much smaller quantities. For example, in the laboratory, it is possible to detect quantities of cesium-137 in soil as low as 1 part in 10^{15}, which is one part in 1000,000,000,000,000.* This detection sensitivity is like picking a single grain of rice from a pile equal to the entire annual production of rice in the United States—some 20 billion pounds of rice harvested from 3 million acres of land.† This is how sensitive radiation detection equipment in a laboratory can be and the reason why radioactive isotopes have been used for decades in research laboratories as "tracer" elements.

UNITS OF RADIATION

When it comes to measuring radiation, there is some bad news and some good news. The bad news is that dozens of radiation units have been defined over the years. A number of these units define a measurable physical quantity, and some units have been created to serve an operational, technical, or regulatory purpose. Some radiation units are no longer in use. Some units have changed their meaning or their names over the years. With increasing use of radiation in medical applications, more complex radiation quantities are likely to be defined in the future. The myriad of units and quantities presents a challenging communication issue for anyone interested in learning about radiation. To make things even more confusing, the international community uses a set of units different from those used in the United States.

The good news is that you do not have to learn them all to grasp basic concepts or benefit from this book. Some understanding of the most fundamental and common units of radiation measurement is necessary and provided here. However, you do not need to commit this information to memory as you read the rest of the book. Just reading over the material and some familiarity with the terms is all that you need.

Radiation units are generally used to describe one of the following properties or quantities (Table 3.3):

- How much radioactivity is present?
- What are the radiation levels?
- What is the radiation dose?

---

* In radiation units, this quantity is equal to 0.1 picocurie of cesium-137 per gram of soil. The unit of curie is described later in this chapter.

† The rice production data are from the U.S. Department of Agriculture, crop production 2007 summary (available from www.usda.gov). This assumes 29,000 grains of rice (long-grain white rice) per pound (taken from www.producersrice.com/rice/facts.html).

Table 3.3 Common Units of Measurement

	Units		
Quantity	**International**	**United States**	**Conversion**
Amount of radioactivity	Becquerel (Bq)	Curie (Ci)	1 Bq = 0.000000000027 Ci 1 Ci = 37 billion Bq
	Mega-becquerel[a] (MBq)	Milli-curie[a] (mCi)	1 MBq = 0.027 mCi
Radiation levels or dose rates	Coulombs per kilogram of air per hour (C/kg/h)[b]	Roentgen per hour (R/h)	For practical purposes, assume 1 R/h = 1 rem/h (1 mR/h = 1 mrem/h).
	Milli-sieverts per hour[a] (mSv/h)	Milli-rem per hour[a] (mrem/h)	1 mSv/h = 100 mrem/h
Radiation dose	Gray (Gy)	Rad	1 Gy = 100 rad
	Sievert (Sv)	Rem	1 Sv = 100 rem
	Milli-sievert[a] (mSv)	Milli-rem[a] (mrem)	1 mSv = 100 mrem

[a] Hyphenation is used in the table only to aid in pronunciation.
[b] This unit is used in scientific literature, but it has no practical utility: 1 C/kg/h = 3,876 R/h.

A "hot" source does not necessarily mean high radiation levels. If the source is an alpha emitter and there is a lot of it, it is a hot alpha-particle source. But the radiation level standing 1 foot from the source is no different than it would be without the source because the alpha particles cannot reach that far. If the source is a gamma emitter, it could be shielded in some lead container and radiation levels in the room may not be high. Furthermore, high radiation levels in a room do not necessarily mean that someone who enters the room for a brief period of time receives a high dose. Thus, separate quantities are needed to measure radioactivity, radiation levels, and radiation dose.

Radioactivity

Units of radioactivity describe the number of radioactive transitions that take place per second. Obviously, the higher the number is, the "hotter" is the radioactive source. The international unit for this quantity is becquerel (pronounced be-kə-rel; symbol = Bq).* One becquerel means one radioactive

* The unit is named in honor of French scientist Henri Becquerel, who discovered the natural spontaneous process of radioactivity in 1896.

transition per second. Most people may not realize that 1 Bq is a very small quantity of radioactive material. For example, an average size banana contains 14 Bq of activity due to its natural potassium content. An adult human body contains about 5,000 Bq of potassium-40 activity. When a real radiation source, rather than bananas, is described, units such as gigabecquerel (GBq), which is equal to one billion Bq, are likely to be used.

An older unit of measure for the amount of radioactivity—the curie (abbreviated Ci)—is still widely used in the United States.* The unit was originally devised to describe the radioactivity levels in one gram (1 g) of radium. Even though it does not appear to be large, 1 Ci of radioactivity is actually a large quantity: 1 Ci = 37 billion Bq or 37 GBq.

To give you an idea of how different the curie and becquerel units are, imagine that they are units of distance. If 1 Ci represents a distance equal to the Earth's circumference (40,000 km or 24,900 miles), then 1 Bq will be equal to 1 mm (just over 1/32 of an inch).

Other variations of the unit curie are often seen. For example, millicurie (mCi) is 1/1,000 of a curie. A common unit used in environmental samples of soil and water is a picocurie (pCi), which is equal to one trillionth of a curie—smaller than a becquerel. This again illustrates the sensitivity of radiation detection equipment in a laboratory.

In the Earth's circumference analogy just used, one picocurie would represent the diameter of a human cell. That is why it is very important to pay attention to the prefixes that come with radiation quantities and appreciate the wide range of values they represent (Table 3.4).

Another common unit you may read or hear is "counts per minute" or its abbreviated form, "cpm." This unit is used to describe the response of a portable radiation instrument when it is used to measure radioactivity levels on a contaminated surface or person. This unit describes how many radiation events are registered (or counted) in the instrument each minute. Unlike becquerels or curies, the unit of cpm is not an absolute quantity. Its value is as dependent on the type of detector used as it is on the amount of contamination present.

Radiation Levels

A common unit to measure gamma-ray or x-ray radiation levels is the roentgen, with the symbol R.† Radiation levels are always expressed per unit of

---

* This unit is named in honor of French scientists Pierre and Marie Curie, who discovered the elements radium and polonium in 1898.

† This unit is named in honor of German scientist Wilhelm Conrad Roentgen, who discovered x-rays in 1895.

Table 3.4 Range of Radioactive Source Quantities

	Amount of Radioactivity[a]	
Source	**Becquerel (Bq)**	**Curie (Ci)**
Daily intake of natural uranium in a typical human diet	0.03	0.0000000000008 (0.8 pCi)
Natural uranium per kilogram of granite rock	50	0.00000000135 (1.4 nCi)
Potassium-40 content in 1 kg of spinach	240	0.0000000065 (6.5 nCi)
Cesium-137 source used in training classes for teaching	3,700	0.0000001 (0.1 μCi)
Amount of cesium-137 exempt from licensing	370,000 (0.37 MBq)	0.0001 (0.1 mCi)
Iodine-125 permanent seed implant to treat prostate cancer	10,000,000 (10 MBq)	0.027 (27 mCi)
Cesium cancer therapy unit in hospital	100,000,000,000,000 (100 TBq)	2,700
Amount of cesium-137 released from the Chernobyl accident	85,000,000,000,000,000 (85 PBq)	2,300,000

[a] When the numbers are too small or too large, a unit prefix is typically used. Examples are given in parentheses.

time—for example, roentgen per hour (R/h). Under routine circumstances, 1 R/h is considered a high-radiation area. Typical natural background radiation levels are in the order of 10 microroentgens per hour (μR/h; 1 μR/h = 0.000001 R/h). In occupational settings, areas that exceed 2 milliroentgens per hour (mR/h) are considered "controlled" areas and access is limited. In an emergency, if the radiation level is at 1 R/h, responders are allowed to work up to 5 hours in that area. If the radiation level is 10 R/h, workers are allowed to work in that area for only half an hour. For life-saving activities, higher radiation dose limits may be authorized (see Chapter 14).

It is also possible to report radiation levels in units of counts per minute. As stated earlier, the unit of cpm does not describe an absolute radiation quantity. Its value depends on the type of instrument used. It is often helpful to relate the reported value to the natural background levels measured with the same instruments. This means that it helps to express radiation levels in multiples of natural background radiation.

> *Example:* Suppose someone reports to you that radiation levels are 3,000 cpm. By itself, that value is not informative. You are also told that a 1-inch sodium iodide gamma scintillation detector was used to make

> that measurement. If you do not know that particular type of instrument, the information is still not helpful. But if you ask about the natural background radiation level measured with the same instrument, you may be told that it is 1,500 cpm. This means that the reported radiation level of 3,000 cpm is twice the natural background radiation.

Ionization chambers measure the intensity of gamma-ray or x-ray radiation in units of roentgens per hour, milliroentgens per hour, or microroentgens per hour. They are the instrument of choice to report gamma radiation levels.

Radiation Dose

So far, we have described radioactive materials or the radiation levels they generate. With the concept of dose, your own exposure levels are introduced into the picture. Radiation dose defines the amount of radiation energy that is imparted, or given up, in your body. The international community uses the units of gray (Gy) and sievert (Sv).* In the United States, the older units of "rad" and "rem" are still widely used (see Table 3.3). The physical description of these units is beyond the scope of this book, but in short, the units of gray and rad are purely physical; they are a measure of radiation energy delivered to a given mass of material including human tissue. The units of sievert and rem, on the other hand, take into account the biological ramifications of that radiation energy.

Because 1 Sv (100 rem) is a relatively large dose of radiation, often the unit of millisievert (mSv) is used to designate radiation doses for occupational practices, medical procedures, or environmental doses. In the United States, the unit of millirem (mrem) is used. For example, the annual limit of radiation exposure for the general public in the United States is 0.001 Sv or 1 mSv (0.1 rem or 100 mrem). This does not mean that any radiation dose higher than this "limit" is necessarily detrimental. The annual limit of radiation exposure for occupational workers in the United States is 0.05 Sv or 50 mSv (5 rem or 5,000 mrem), and radiation dose from an average whole-body CT scan is 0.01 Sv or 10 mSv (1 rem or 1,000 mrem).

Dose Rate If a radioactive source is nearby or an area is contaminated, it is customary to report the radiation dose value per unit of time. For example, if it is stated that radiation dose rate at a specific location is 0.2 mSv per hour

* These units are named in honor of Luis Harold Gray, the British scientist considered to be the father of radiation biology, and Rolf Sievert, a Swedish scientist with major contributions to the field of medical physics and radiation biology.

(20 mrem per hour), that means people staying at that location for 1 hour will receive a total dose of 0.2 mSv (20 mrem). If people stay there for 3 hours, they receive three times this dose—that is, 0.6 mSv (60 mrem). As Chapter 6 will show, dose rate is an important consideration when the health effects of radiation exposure are examined.

Committed Dose When the source of radiation is outside the body and you are standing close enough to it to be irradiated, you are receiving a radiation dose. As soon as you walk away, irradiation stops and there is no additional dose. In contrast, when the radioactive material is inside the body, it will continue to irradiate the body from the inside as long as it remains there. The term *committed dose* is used in these circumstances to describe the radiation dose that the radioactive material *will* deliver over the next few days, months, or years.

In other words, committed dose describes a dose that has not been received in full yet! How long the radioactive material remains inside the body depends on the half-life of the material, the body metabolism, and whether any drugs are available to help the body get rid of that radioactive material faster. If treatment is available and provided, the committed radiation dose can be reduced but not entirely eliminated. This topic is the subject of Chapter 12, "Radiation Drugs."

Quite often, the committed dose is calculated for 50 or 70 years into the future. In many cases, most of the committed dose is absorbed by the body within the first year, and none or little is left for the remaining years.

> *Did you know?* In occupational settings, some radiation workers have accidentally inhaled radioactive materials with long physical and biological half-lives. In these cases, the radioactive material deposits in the person's bones and some of that material stays there for decades, delivering a small quantity of radiation dose during the entire lifetime. In contrast, for nuclear medicine procedures, radionuclides with short half-lives are injected into patients for imaging purposes, and all of the radioactive material decays and leaves the patient's body in a matter of days or weeks.

Now that you have gotten through the most difficult chapter in this book, let us finish with two thoughts:

- If radiation concepts, quantities, and units cause you confusion, you are not alone and you should not be discouraged.
- If you are ever in a situation dealing with radiation or radioactivity and someone communicates to you using numbers and radiation units, *demand* that the quantities be explained to you and the values be placed in some familiar context so that you and everyone

else can gauge the magnitude of those measurements. It is helpful to know the basic concepts explained in this chapter, but it is not your job or responsibility to know these radiation units or quantities.

Resources

A number of online resources can supplement the information you learned in this chapter:

The U.S. Environmental Protection Agency Web site for students and teachers is an excellent resource (http://www.epa.gov/rpdweb00/students.html).

The Radiation Event Medical Management (REMM) Web site has helpful animations and diagrams to illustrate basic radiation concepts (http://www.remm.nlm.gov/imagegallery.htm).

The Centers for Disease Control and Prevention offers an extensive glossary of radiation terms (http://emergency.cdc.gov/radiation/glossary.asp).

The University of Michigan Health Physics Web site (http://www.umich.edu/~radinfo/introduction/index.htm) and the Idaho State University Radiation Information Network (http://www.physics.isu.edu/radinf/) have informative pages on radiation principles.

For more detailed descriptions of radiation concepts, quantities, and units; mathematical relationships; and more specialized topics, any one of the following textbooks will be helpful:

Bevelacqua, J. J. 2004. *Basic health physics, problems and solutions.* Weinheim, Germany: Wiley–VCH.

Cember, H. 1996. *Introduction to health physics,* 3rd ed. New York: McGraw–Hill.

Gollnick, D. A. 2004. *Basic radiation protection technology,* 4th ed. Altadena, CA: Pacific Radiation Corporation.

Turner, J. E. 2004. *Atoms, radiation, and radiation protection,* 2nd ed. Weinheim, Germany: Wiley–VCH.

Part Two

THE IMPACT OF RADIATION

4

ACCIDENTAL RADIATION EMERGENCIES

OVERVIEW

Radiation control authorities worldwide have always been concerned with safe use and handling of radioactive materials and radiation-generating devices because of their wide applications in medical, industrial, and research environments as well as in power generation. In the last 100 years, literally hundreds of radiation accidents have been documented. Many more have occurred, but without formal documentation. In terms of ability to cause physical damage or threat to human health and safety, radiation accidents have a wide range of consequences, from minimal to catastrophic.

The great majority of radiation incidents have been inconsequential, with little or no potential risk to human health; these types of incidents occur routinely around the world. Dozens of radiation accidents have caused serious injury or death to one or more individuals. Only a few radiation accidents have had catastrophic consequences.

> *Terminology alert:* In some contexts, there may be a distinction in the meaning of the words "event," "incident," "accident," or "emergency." Exact definition and usage of these words may change with time as they have in the past. In our discussion, these words will be used interchangeably. In some instances, exact wording is important. For example, in discussing nuclear power plant accidents, the term "site area emergency" conveys a specific meaning, which we will describe. In this context, the word "emergency" cannot be substituted with another.

One important classification that we will use right now for our discussion is whether a radiation incident is caused accidentally or intentionally. In this chapter, we will review various types of radiation emergencies that can happen accidentally. In the next chapter, we will review the types of radiation emergencies that can result from intentional action, including terrorism-related incidents.

As we describe these types of accidental radiation emergencies, you will learn the potential impacts of each using real examples such as the Three Mile Island (TMI), Chernobyl, and Goiânia accidents, as well as a number of lesser known incidents. The potential health impacts of radiation emergencies are described only in a general way here; we save the more detailed description for Chapter 6. Later, in Chapter 9, we describe the steps you can take to protect yourself in various types of radiation emergencies.

There is probably more information in this chapter than you need. If you are interested only in terrorism-related incidents, you can skip ahead to the next chapter. However, it is recommended to at least skim over this information. Understanding the ranges and types of radiation emergencies, together with the basic information already reviewed in Chapters 2 and 3, will help you in applying the information we cover in later chapters.

The unintentional, accidental types of emergencies we describe here can happen potentially at any place where radioactive sources or radiation-generating devices exist or are used. These sites include hospitals, research laboratories, roadways, construction sites, smelting furnaces, nuclear power plants, and in space. Please note that the purpose of this information is not to alarm you, but rather to demonstrate that we live with a baseline of risk for radiation accidents in modern society. On the positive side, these types of accidents have improved our ability to respond to them, resulting in continuous improvements in safeguards and encouraging safer use of radiation and radioactive materials.

LOST OR ORPHAN SOURCES

"Orphan sources" refers to sealed (self-contained) radioactive sources that are no longer in the custody of the original owner and may not be under proper control. This presents a major challenge to radiation control authorities around the world. In the United States alone, there are more than 150,000 licenses authorizing the use of radioactive materials and approximately 2 million radioactive material-containing devices that are distributed under general license. This figure does not include items such as smoke detectors, which can be distributed to persons exempt from licensing.* You may have already guessed that these radioactive sources vary greatly in terms of type, quantity, and the danger they may pose should they be lost, misplaced, or stolen.

In the United States, as many as 500,000 devices that contain some amount of radioactive material are no longer needed. Proper disposal of these sources

* Smoke detectors contain a small amount of americium-241 for their proper functioning. This does not present a hazard to homeowners.

is an issue. By far, the majority do not present a health risk. However, if the sources are lost and end up in a smelting furnace, for example, the incident can still cause an economic loss to the facility owner.

However, some types of radioactive sources can cause significant injury—even death—if they are misplaced and end up in the hands of someone with no knowledge of their radioactivity. We will discuss a few examples of such cases later in this chapter. There is also the risk of intentional, criminal use of these sources to inflict harm. The criminal intent could be in the form of a "dirty bomb" that terrorists may use or in other forms as described in the next chapter. Examples of high-risk sources include high-level radioactive sources used to power lighthouses and navigational beacons in remote areas of the former Soviet Union, a number of which have been vandalized to salvage the metal, and radioactive sources that were once used as seed-irradiators (Figures 4.1 and 4.2).

Figure 4.1
These mobile cesium irradiators were used in the 1970s for agricultural research in the former Soviet Union, under a project name, "Gamma Kolos," to study the effects of radiation on plants and animals. The canisters, each containing 3,500 Ci of cesium-137, are among the few that have been recovered in Georgia and Maldova. An unknown number remain unaccounted for in the former Soviet Union. (Photo: IAEA.)

Figure 4.2
A team member tests his handheld radiation detector during a June 2002 survey in Georgia searching for orphan sources. (Photo: Petr Pavlicek/IAEA.)

Therefore, controlling these sources has become a priority on an international scale with many countries working together through the International Atomic Energy Agency (IAEA) to improve the security of radioactive sources. Because securing all radioactive sources is not practical, a classification system is used by the international community to prioritize the radioactive sources in common use into five different categories (Table 4.1). This categorization is based on the level of danger the source presents in terms of causing radiation injury; category 1 is the most dangerous and category 5 the least dangerous type of radioactive source material.

The most serious radiation accident involving an abandoned or orphan radioactive source occurred in 1987 in Goiânia, Brazil, when a radiotherapy source was left in an abandoned clinic. This source was an IAEA category 1 source (most dangerous). We describe the Goiânia accident later in this section.

Misplaced radium needles also belong to the category of lost or orphan sources. These devices were used for medical procedures by a variety of ophthalmologists, gynecologists, dermatologists, radiologists, or even nonphysicians at a time when radium was the only source for radiation therapy; such

Table 4.1 Categories of Radioactive Sources Used in Common Practice

	Description of Hazard[a]	
Category	**Exposure to Unshielded Source**	**Dispersion of Source Due to Explosion or Fire**
1—Extremely dangerous	Likely to cause permanent injury within a few minutes, probably fatal if exposure lasts for a few minutes to an hour	Permanent injury or death possible (although unlikely) in immediate vicinity; little or no risk of immediate health effects beyond a few hundred meters
	Examples: industrial irradiators; teletherapy sources; radioisotope thermoelectric generators (RTGs)	
2—Very dangerous	Could cause permanent injury within minutes to hours, possibly fatal if exposure lasts for hours to days	Permanent injury or death possible (although very unlikely) in immediate vicinity; little or no risk of immediate health effects beyond 100 m
	Examples: industrial radiography sources; high/medium dose rate brachytherapy sources	
3—Dangerous	Could cause permanent injury within hours, possibly (although unlikely) fatal if exposure lasts for days to weeks	Permanent injury or death possible (although extremely unlikely) in immediate vicinity; little or no risk of immediate health effects beyond a few meters
	Examples: fixed industrial gauges with high-activity sources such as well logging gauges	
4—Unlikely to be dangerous	Unlikely to cause permanent injury, possibly (although unlikely) to cause temporary injury	No permanent injury[a]
	Examples: low-dose-rate brachytherapy sources; moisture/density gauges; bone densitometers; static eliminators; medical unsealed sources, I-131	
5—Most unlikely to be dangerous	No permanent injury[a]	No permanent injury[a]
	Examples: positron emission tomography (PET) check sources; other low-activity, nonexempt sources	

Source: Adapted from IAEA. 2005. Categorization of radioactive sources. Safety guide no. RS-G-1.9. Vienna, Austria.

[a] This IAEA hazard analysis is primarily concerned with immediate health effects, injuries, or possibly death. Possible long-term health effects (such as cancer) are a secondary concern in this context. Exposure to higher category sources would represent a higher risk of cancer.

procedures did not necessarily take place under supervision of authorized radiation oncologists. As other devices and radionuclides became available, many radium needles were donated or simply abandoned. Because many of the original owners have passed on, it is not unusual to find radium needles in safes, bank vaults, storage chests, and home garages, or recycled into jewelry. Radium needles present a radiation exposure hazard and prolonged exposure can cause injury.

It is also not unusual for hospitals or clinics to lose or misplace radioactive sources. Every year, the Nuclear Regulatory Commission (NRC) receives approximately 200 reports of lost, stolen, or abandoned radioactive sources and devices. Many such cases also go unreported because the owners may not be aware that the source is missing or that they are required to report the loss. Many of these sources contain low amounts of radioactivity and present little or no safety hazards.

The publicly available NRC event notification reports were monitored for a period of 2 months. During that period, several reports of radioactive sources lost in health care facilities were submitted. These included an iridium-192 brachytherapy seed, iodine-125 seeds, strontium-90 eye applicator, iodine-131 inorganic salt, technetium-99 syringe, sulphur-35, and tritium used for research. The explanation of how they were lost included "likely removed by the cleaning crew," "discarded in trash," "washed down the drain to the public sewer system," "could not locate," "misplaced," or "missing."* Given the large number of sources in use at any given time, the frequency of these incidents is perhaps not unusual. Fortunately, however, the great majority of these incidents is largely inconsequential and does not pose a health risk to the public because of a low amount of activity, often short half-lives, and a low probability of coming in contact with people.

Case Study: Goiânia, Brazil

This unfortunate accident is undoubtedly the worst accident on record involving a radioactive source. A team of two doctors moved to a new location and left behind an unneeded teletherapy unit in the clinic building, which was subsequently abandoned. The teletherapy unit contained the commonly used source cesium-137. The radioactive source was secured and shielded in position, but the building was not secured. Two local men entered the clinic and dismantled the teletherapy unit, thinking the piece of equipment might have some value as salvage metal. They removed pieces, including the source wheel, but in the process punctured the source capsule with a screwdriver

* The NRC event notification reports can be accessed at www.nrc.gov/reading-rm/doc-collections/event-status/event/

Figure 4.3

A scrap yard at the Rua 6 (6th Street) neighborhood in Goiânia, Brazil, where the tragic 1987 accident occurred. This picture was taken in October 2007, 20 years after the accident. (Photo: Kirstie Hansen/IAEA.)

and some of the radioactive material spilled. The source was in the form of cesium-chloride salt that was the size of rice grains, which could be readily crumbled into powder.

Some pieces, including the source capsule, were sold to a junkyard owner (Figure 4.3). The owner and his wife noticed a "blue glow" emanating from the source in their garage and, thinking that the powder might be valuable or even have supernatural properties, took the source inside their home. Later, some of the source material was shared with their friends and families. Some people rubbed the powder on their skin. A 6-year-old girl ate after handling the radioactive material, thus ingesting significant quantities of the source. Over the next several days, as people started developing symptoms of radiation sickness, the junkyard owner's wife took what was left of the source and transported it by bus to a hospital. They first suspected some tropical disease until one alert physician suspected radiation injury and a radiation physicist detected the radioactivity; soon afterward, the Brazilian radiation authority responded to the incident.

The town of Goiânia was not in rural Brazil. It was a thriving metropolitan area with a population of nearly one million people. In the aftermath of this incident, 112,000 people were monitored for radioactive contamination. Of these, 249 people were found to have some radioactive contamination, 120 had only external contamination, and 149 also had internal contamination (see Chapter 3 for definitions of these terms). Fifty-four people were hospitalized and eight developed acute radiation syndrome (see Chapter 6 for a description of radiation sickness). Within a few weeks, four people died, including the 6-year-old girl, who had the highest level of internal contamination. The junkyard owner's wife also died.

Five years later, the junkyard owner was the fifth person to die. Several people had severe skin injuries and one person's forearm was amputated. Forty-six homes were found to be contaminated, and when decontamination activities were completed, 3,500 m^3 (4,600 cubic yards) of waste had been accumulated and stored in more than 5,000 different containers. The intact cesium chloride source in the capsule had weighed only 90 g (3.2 ounces). Given the high level of radioactivity involved and the fact that it took 3 weeks for the accident to be recognized, it is remarkable that the extent of damage was so limited. This accident, as tragic as it was, could have been much worse (Figure 4.4).

The Goiânia accident had many psychosocial implications (see Chapter 8), some lasting to the present day. The accident has provided many lessons and a wealth of knowledge about the consequences of radiological accidents, including the treatment of internal contamination with radioactive cesium (see Chapter 12).

Case Study: Samut Prakan, Thailand

Twelve years after Goiânia, an accident occurred in Thailand that had an eerie resemblance to the Goiânia accident. In the fall of 1999, a company in Bangkok transferred one of its teletherapy units from a warehouse to an unsecured storage location without obtaining authorization or informing the Thai radiation control authorities. In January 2000, several individuals obtained access to the building, partially disassembled the unit, and took the parts to the home of one of the individuals; there, they continued to disassemble the unit. The teletherapy head had the trefoil radiation sign and a warning label. However, none of the individuals recognized the sign for radioactive material and the warning label was not in their native language.

The partially disassembled unit was subsequently taken to a junkyard in Samut Prakan (approximately 30 km southeast of Bangkok) to be sold. At the junkyard, while a torch was used to further disassemble the unit, the radioactive source capsule containing cobalt-60 fell out without being noticed (Figure 4.5). By the middle of February, several individuals had gotten ill and when they sought medical help, an alert physician recognized the

Figure 4.4

The Rua 6 (6th Street) neighborhood in Goiânia, Brazil, 20 years after the accident. Lessons learned from the Goiânia accident continue to shape radiation safety and emergency response planning today. (Photo: Kirstie Hansen/IAEA.)

signs and suspected that a radiation source was involved. Three individuals died of radiation injury.

It was fortunate that, unlike the case in Goiânia, the radioactive source capsule had not been breached and no radioactive contamination occurred. Therefore, radiation injury was due to irradiation only, and no dispersal of radioactive material took place, thus limiting the extent of the accident.

This accident spurred international efforts to design a new international symbol to communicate hazards from high dose-rate radioactive sources better. You will find details about the new international sign in Chapter 9.

MEDICAL ACCIDENTS

In Chapter 2, you read about widespread use of radiation in medicine. The more serious medical accidents can result from overexposure to radiation-generating devices. Nearly always, the problem is due to human error and

Figure 4.5
Cylindrical pieces that fell apart during disassembly of the teletherapy unit at a junkyard in Samut Prakan, Thailand, in January 2000. The cobalt-60 source was contained in this assembly. Three individuals died of radiation injury. (Photo: IAEA.)

a failure to follow proper procedures. Overexposure to a radiation-generating device can occur during radiotherapy treatment and, if the dose is high enough, may cause death.

One such incident occurred at the San Juan de Dios Hospital in San José, Costa Rica, in August 1996, when a cobalt-60 radiation therapy source was replaced with a new one. When the new source was calibrated, the person in charge of dosimetry made a calculation error that resulted in the administration of higher radiation doses to patients. The mistake was discovered a month later and seven deaths were attributed to this overexposure. A few of the surviving patients suffered debilitating and severe radiation injuries as a result of this accident.

Another example occurred at the Instituto Oncológico Nacional (ION) in Panama City, Panama, when 28 cancer patients were overexposed between August 2000 and March 2001. The error occurred when a computerized treatment planning system changed and technologists used the modified protocol without a verification test. In spite of the treatment times being about twice those required for correct treatment, the error went unnoticed for some time. Five of the early patient deaths and several later deaths were attributed to this radiation overexposure.

Overexposures can also occur with linear accelerators, which provide another method of delivering radiation therapy. Nearly always, these types of accidents are caused by human error. A tragic example of such an accident occurred in late 1990 and early 1991 in a radiotherapy clinic in Zaragoza Clinical University, Zaragoza, Spain. In that facility, a linear accelerator delivered radiation therapy to patients. However, a miscommunication between the maintenance crew and the physicists in charge of calibration, coupled with a failure to follow procedures for periodic verification of the radiation beam (human error), resulted in overexposure of 27 patients, 18 of whom died as a result of overdoses that ranged from 200 to 700%.

Overexposure can also occur during lengthy fluoroscopic procedures. Fluoroscopy uses x-rays to obtain real-time, live, moving pictures of internal organs as the attending physician or surgeon is performing a procedure. Coronary angioplasty is one example. Above a certain dose threshold, radiation can cause damage to the skin, reddening and skin rash, hair loss, and, in extreme cases, necrosis of tissue. One example of such overexposure occurred in the United States in March 1990, when a patient underwent a coronary angiography, coronary angioplasty, a second angiography procedure due to complications, and a coronary artery bypass graft on the same day. In the weeks and months that followed, skin on the back of the patient where the fluoroscopy beam had been directed became progressively worse, starting from red skin to looking like a burn and then severe tissue necrosis. This injury eventually required a skin graft.

Medical accidents can also result from brachytherapy (seed implantation) procedures. This procedure is used to treat a number of different cancers, such as prostate or cervical cancer. The technique involves carefully placing the radioactive "seeds" inside or next to the cancerous tissue to irradiate this tissue. The seeds, which are very small (the size of a grain of rice), are then carefully taken out of the body.

In 1992, an elderly woman suffering from cancer in Indiana City, Pennsylvania, received brachytherapy treatment. The sliver of radioactive seed (iridium-192) became detached accidentally as the source was retracted. This condition could have been easily noticed before the patient was released, but due to human error, it was not. The woman returned to her nursing home with the source inside her. The source, together with surgical dressings, eventually became dislodged from the patient and was discarded as medical waste. The patient died a few days later due to injury from the radiation overdose. A week later, the waste was discovered when it tripped an alarm at an Ohio waste disposal facility. Ninety members of the public, including people at the nursing home, were accidentally exposed to the radioactive source.

Medical overexposures can also result from misadministration of therapeutic drugs that contain radioactive materials. When these incidents occur, the overexposures are typically not significant.

INDUSTRIAL ACCIDENTS

One of the common accidents involves industrial radiography sources. These high-dose-rate devices are used for nondestructive testing of equipment and materials (such as welds and piping) by producing an image of that material on radiographic film—almost exactly the same process used to produce an x-ray image. Industrial radiography sources are used in cases when the object is too thick for conventional x-rays or is shaped in such a way or confined in an area where portable x-ray machines cannot be used. The imaging system containing these sources is sometimes referred to as a camera.

Typically, the radioactive sources are cesium-137, cobalt-60, or iridium-192, which are usually shaped as small metal capsules at one end of a short flexible cable called a "pigtail." The other end of the pigtail is connected to a long cable that the radiographer uses to operate the source. Accidents occur when the pigtail detaches from the crank-out cable and is left behind at the job site without the operator noticing (operator error). The small shiny capsule with no marking or warning sign may then be picked up by another person, who will be in danger of severe radiation injury leading to limb amputation or even death. Loss of these sources has resulted in many injuries and a number of deaths around the world.

In June 1979, an industrial radiography worker in Los Angeles, California, failed to secure the source at his job site before leaving. The source fell out of the camera and was picked up by another worker, who was not aware that it was radioactive and unshielded. The worker put the source in his back pocket and left it there for 45 minutes before giving it to the plant manager. That length of exposure was enough to cause severe injury to the worker's buttocks. Four other workers suffered radiation injury to their hands. Interestingly, in terms of causing physical injury to people, this accident was far more severe than the accident at the Three Mile Island nuclear power plant just 3 months earlier.

Similar industrial radiography accidents have occurred in many countries around the world, including Chile, South Africa, Iran, Russia, and China. A more tragic case involving fatalities occurred in Mit Halfa, Al-Qalyubiah, just north of Cairo, Egypt. A radiography source used for checking welds in a gas pipeline was lost and recovered by a man who took it home, thinking that it had some value. A month later, his 9-year-old son became ill and died. The man, his wife, and four more children became ill and the father also died before authorities became aware that the family was suffering from radiation sickness due to exposure to the source. Nearly 200 friends and neighbors received some dose of radiation from this source before it was recovered.

Density-moisture gauges are another source of radioactivity whose use is widespread. They are used to determine density of asphalt, soil, aggregate, or concrete or to determine the moisture content of soil or aggregate. The sources are typically cesium-137 (for checking density) and americium-241/beryllium (for checking moisture). The equipment is inherently safe to operate and does not require much training. Unfortunately, it is not uncommon for this equipment to be left behind on the side of a road.

In petroleum exploration, similarly designed equipment is used for well logging operations to evaluate subsurface geological parameters. The operators may leave this equipment unattended or unsecured. The equipment can also fall off the truck. Frequently, a vehicle or pickup truck is stolen with well-logging equipment firmly secured but left in the truck bed. Most often, these sources, which present no value to the car thief, get tossed aside and are lost. They generally pose no significant health risk as long as they remain in their locked and secured shielded casing.

Another type of industrial accident occurs occasionally at metal recycling facilities when radioactive sources that have been lost, stolen, or abandoned get mixed with scrap metal for recycling and end up in a smelting furnace. From 1983 to 1999, 33 such incidents were reported in the United States—mostly at steel mill facilities, but also at facilities that melt aluminum, copper, gold, zinc, or lead scrap. Significant improvements have been made in recent years by installing radiation detectors at metal recycling facilities to examine incoming shipments.

These smelting accidents usually affect the facility and are costly in terms of decontamination and downtime. However, they are unlikely to present any safety threat outside the facility. Radiation doses to mill workers and the public from reported incidents have been low and below regulatory limits.

Similar accidental smelting has been reported in other countries, including Canada, Mexico, Brazil, Italy, Ireland, Sweden, Austria, Spain, Greece, Poland, Bulgaria, Czech Republic, Russia, Kazakhstan, South Africa, China, and India. In a number of these cases, the smelted and then contaminated product (usually steel) was subsequently exported to the United States before it was detected.

TRANSPORTATION ACCIDENTS

In the United States alone, millions of packages of radioactive material are shipped annually by rail, air, sea, and road. These packages contain small quantities of radioactive material that are typically used in research, industry, and medicine. Regulations limit the amount of radioactivity and packing requirements designate how they are shipped. This ensures that if a package is damaged during transport, any contents that are released will not cause substantial health risks. Transportation of spent nuclear fuel or high-level radioactive waste requires specialized casks with stringent safety standards to withstand impact, puncture, fire, and water immersion while maintaining the integrity of the content. This packaging undergoes extensive testing; a number of interesting photos and videos of such tests are available from http://www.sandia.gov/tp/SAFE_RAM/SEVERITY.HTM.

If a transportation accident occurs, the consequences most likely would be confined to the local and immediate surroundings. A potential for exposure to radiation and radioactive contamination exists, but the overall impact is minimal and limited. If the source is released during an explosion or fire, some of the radioactivity would be dispersed to some distance; however, the extent of release would be limited and health risks would not likely be significant.

SPACE MISSIONS

For many years, radioactive materials have enabled extraordinary space missions such as the *Voyager* spacecrafts, the *Galileo* spacecraft's mission to Jupiter, and the Cassini–Huygens mission to Saturn. For these long-term missions away from the Sun, neither fuel cells nor solar energy can provide a power source. However, the energy from radioactive decay of plutonium can. Two types of such plutonium devices can be used: radioisotope thermoelectric generators (RTGs) and radioisotope heater units (RHUs).

RTGs use the heat energy from radioactive decay of plutonium-238 and convert that energy to electricity using a thermoelectric converter. The electricity powers various components of a spacecraft, including computers and communication devices. This is a reliable, long-lasting source of electrical power that is not affected by the extreme cold temperatures of space. RHUs contain much smaller amounts of plutonium dioxide and are used to generate heat and regulate the temperature for certain instruments and the inside of the spacecraft. The Cassini mission, for example, had three RTGs and 129 RHUs on board.

The RTG units have been carefully engineered and constructed to survive a range of possible accidents, such as explosion or fire at launch, inadvertent reentry into the atmosphere during a gravity-assisted swing-by, and impact on land or sea. Nevertheless, each time an RTG-powered spacecraft prepares to launch, teams of experts are deployed to the area. In case an accident occurs during launch, the expert responders are already on the scene to determine quickly whether any release of radioactive material from RTGs or RHUs onboard the spacecraft has occurred. The probability of these accidents is low. In addition, environmental impact studies have shown that even if such an accident occurs, the chances of any adverse health effects are extremely low.

The *Apollo 13* lunar mission contained an RTG. When the spacecraft was forced to return to Earth, the lunar module was jettisoned as it approached and burned up during reentry. The RTG survived the reentry into Earth's atmosphere, as designed, and plunged into the Pacific Ocean intact with no released contamination, as confirmed by air sampling.

When disaster struck space shuttle *Columbia* in 2003, there was no RTG onboard, but the shuttle contained numerous smoke detectors (containing americium-241) and several research radioisotopes in low quantities. This equipment presented no radiological hazard; nevertheless, the debris was examined for traces of radioactivity. Authorities issued a notice warning people who might identify and come in contact with the shuttle debris not to touch the debris material. The concern was not for radioactivity, but rather for the possibility that small residual amounts of toxic liquid fuel or other toxic and caustic chemicals might have survived the descent.

Not all RTGs are used for space missions or are under careful security controls. Because of their long life and few maintenance requirements, the former Soviet Union used an estimated 1,000 RTGs, mainly to power lighthouses and navigational beacons in remote areas. These high-dose-rate devices use the isotope strontium-90 and many of them are not under government control. Many are no longer maintained and have been abandoned. Vandalism of these devices for scrap metal and leaks of radioactivity into the surrounding area have been reported. The international community has recognized these RTGs as a safety threat not only because people may unknowingly approach these devices, but also because they present a tool that can be used by terrorists in construction of dirty bombs (more on this later).

Did you know? In January 1978, the Soviet spy satellite *Cosmos 954,* powered by an onboard nuclear reactor, started to lose orbit. The reactor failed to separate on command, and the satellite descended through the atmosphere and crashed in Northwest Territories, Canada. A number of highly contaminated pieces of debris were eventually recovered. Fortunately, the territory was remotely located.

NUCLEAR POWER PLANTS

For many people, nuclear power is associated with images of devastation, danger, and fear. These images are unfortunate because nuclear energy today is one of the most tightly regulated industries and perhaps one of the safest. Nuclear power reactors work by using heat from nuclear reactions to convert water to steam, which in turn powers generators to produce electricity. This conversion occurs in a contained environment. It is important to recognize that even if a nuclear power reactor is heavily damaged or sabotaged, all the engineering controls are lost, and the containment structure fails, it is still impossible for a nuclear explosion, such as the ones at Hiroshima and Nagasaki, to occur at that reactor.*

In a nuclear power plant accident, the worst-case scenario affecting the surrounding areas is a breach of the containment structure and release of large amounts of radioactive materials, as occurred at the Chernobyl reactor in Ukraine in 1986. Current design, structural, and operational safety features of nuclear power plants, especially in the United States and other industrially advanced countries, make this worst-case scenario highly unlikely.†

If radioactive materials from a nuclear power plant are released to the environment, this release is characterized as a plume, or cloud, of radioactive particles and gases, many of which are short lived. This type of accident can expose the population in the surrounding area to radiation. Exposure can occur as a result of radioactive materials depositing or settling on skin (contamination), or through inhalation or ingestion of radioactive particles. The radiation doses to people in surrounding communities under these conditions could be significant, but not likely to be life threatening. The two hazards of concern that we discuss in later chapters are inhalation of iodine-131, especially for children, who have a higher risk of developing thyroid cancer later in life, and skin burns due to contamination on the skin if radioactive particles are not washed off within several hours or days.

Many countries with nuclear power reactors have well-developed emergency response plans that are exercised and evaluated on a regular basis. In the United States, 104 commercial nuclear reactors are currently operating at 65 locations in 31 states.‡ If you live within a 10-mile radius of such a

* The type of nuclear fuel used in a power reactor, physical dimensions of that fuel, and how the fuel is packaged make it impossible for a nuclear power reactor to detonate like a nuclear bomb.

† The Chernobyl reactor used graphite as a moderator, which made it vulnerable to loss of water and extreme temperatures. When water is lost with this style of reactor, chain reactions accelerate. In contrast, U.S. commercial power reactors use water as a moderator, which also serves to self-regulate against loss of water and overheating (i.e., when water is lost, chain reactions stop). Furthermore, the containment structure at Chernobyl was not built to withstand the explosion that occurred. Such a containment design could never be licensed in the United States.

‡ It is common for one facility to contain more than one nuclear reactor on site. As of April 2009, 26 new reactors were planned in the United States—many of them at existing sites.

facility, you should be familiar with planned evacuation procedures in your community. Although we will describe these response activities later, it is important to note here how critical it is for the public to listen to and follow the advice and instructions given by emergency response authorities under those circumstances.

Local emergency response plans define an area within a certain radius of the plant where such radiation exposures are possible. In the United States, this area is referred to as the 10-mile emergency planning zone (EPZ).* This zone is also referred to as the *plume exposure pathway* EPZ because primary concern in this zone is to reduce exposure to the plume of radioactive materials that may be released from the reactor.

A second zone, the 50-mile emergency planning zone (EPZ), designates where radioactive material accidentally released from the plant can contaminate water supplies, agricultural crops, livestock, and, consequently, the local milk supply.† This zone is also referred to as the *ingestion exposure pathway* EPZ because primary concern in this zone is to reduce the possibility of internal contamination through ingestion of potentially contaminated food and water.

In principle, response to and evaluation of an accident at a nuclear power plant can be simplified as follows. Does the accident have potential to release radioactive materials to the outside environment? If so, an evaluation of the potential impact is made very quickly. It may be prudent to advise people in the surrounding area to "shelter in place," which means to stay indoors and close doors and windows until the release has terminated and the plume (cloud) of radioactive materials has passed. It may be prudent to advise evacuation of certain areas in the immediate vicinity of the power plant. It may also be prudent to issue advisories to people living as far as 50 miles away to refrain from consuming uncovered food items or crops until further evaluation.

If there is a possibility that people may have been contaminated by radioactive material on their skin or clothing, authorities will provide instructions on how to remove contamination. In addition, authorities should be prepared to screen and assist with decontamination of people who need it. We describe all these actions in more detail in later chapters of this book. A segment of the population, especially children, may be advised to take potassium iodide pills. We describe what you need to know about potassium iodide in Chapter 12.

---

* In the IAEA terminology that many countries use, this zone is referred to as the urgent protective action planning zone (UPZ).

† In IAEA terminology, this zone is referred to as the longer term protective action planning zone (LPZ).

Even in the absence of any significant releases of radioactivity, an accident at a nuclear power plant will naturally cause stress, uncertainty, and mental health issues, which will need to be addressed. Such was the case with the 1979 Three Mile Island accident near Harrisburg, Pennsylvania. Not a single individual was physically hurt in that accident—the worst that has ever happened in the United States. The radioactive materials released from TMI caused negligible and inconsequentially small doses of radiation to the surrounding population.

Before closing this section, let us review two terms you are likely to hear and then review the formal classification of emergencies at a nuclear power plant. In the preceding paragraph, we said that authorities will decide whether it is prudent to advise certain actions to protect the public. These are referred to as "protective action recommendations." One factor that the emergency response authorities consider when issuing such recommendations is, of course, the radiation dose that the public is at risk of receiving. Authorities estimate the projected radiation dose in the absence of any action (i.e., if nothing is done).

If, using cautious assumptions, the projected radiation dose is extremely low, the best option could be to take no action. If the projected radiation dose is estimated to be higher than a set dose limit, then certain protective actions may be warranted. That set dose limit is referred to as a protective action guide (PAG). The PAG represents a projected radiation dose that people would like to avoid. If conditions suggest that the projected radiation doses will reach or exceed the PAG level, authorities recommend protective actions to avoid that dose. In this sense, the PAG serves as a trigger level for decision making. In the description of emergency conditions at nuclear power plants that follows, we refer to these trigger levels, or PAGs.

In the United States, all local, state, and federal authorities follow the PAGs established by the U.S. Environmental Protection Agency.*

* U.S. Environmental Protection Agency. 1992. Manual of protective action guides and protective actions for nuclear incidents. EPA 400-R-92-001. The updated document is available from www.epa.gov/rpdweb00/rert/pags.html

Emergency Classifications

In the United States, emergency conditions at a nuclear power plant are described by four classifications.* These are described next in the order of increasing severity:

Unusual event means that some safety issue has arisen or is about to arise at the facility. However, no release of radioactive material from the plant is expected. Unusual events occur from time to time at operating nuclear power plants. During an unusual event, response organizations are notified to ensure a state of readiness just in case the situation escalates.

Alert means that the safety issue has degraded and there is potential for release of radioactive material off-site from the plant. However, even if this release occurs, the amount of radioactive material released is expected to be much lower than the PAGs. In other words, even if radioactive materials are released to the environment, the levels are expected to be very low. Depending on the underlying cause of the alert, environmental measurements may be conducted to monitor the situation and verify that conditions are as expected; however, many types of alerts do not necessarily trigger environmental monitoring.

Site area emergency means that a major failure of plant functions has occurred or is likely to occur. Release of radioactive material from the plant could occur. If so, such a release could exceed the PAG trigger levels within the plant boundary, but not the off-site or outside facility boundary (fence). In other words, no protective actions for the public are issued, but environmental monitoring teams are dispatched to monitor the environmental conditions, and emergency response personnel are on standby and ready to assist with evacuation just in case the situation escalates.

General emergency is a situation where substantial damage to the reactor core has occurred or is imminent. There is potential for melting of reactor fuel and a potential for loss of containment integrity. In other words, there is a good chance that, if radioactive materials are released from the plant, they will exceed the PAGs outside the

* To communicate the significance of a radiation emergency to the international community (whether it is related to a nuclear reactor or not), the International Nuclear Event Scale (INES) is used. Although the U.S. Nuclear Regulatory Commission uses the INES rating to report applicable events to IAEA, this rating system is not used in the United States (for more information about INES, see www.iaea.org/Publications/Factsheets/English/ines.pdf).

> plant's boundary. Predetermined protective actions are communicated to the public and the response authorities take additional measures as necessary to protect the public.

Case Study: Three Mile Island

The Three Mile Island nuclear power plant, like many others in the United States, contained more than one nuclear power reactor. TMI had two units, referred to as TMI unit 1 (TMI-1) and TMI unit 2 (TMI-2). The two reactors were situated on an island near Middletown, Pennsylvania, 10 miles southeast of Harrisburg (Figure 4.6).

In the early morning hours of March 28, 1979, a series of events began to take place, including equipment malfunctions, design-related problems, and worker errors. The accident ultimately resulted in a partial meltdown of the TMI-2 reactor core, and some radioactive materials were released off-site. Reactor core meltdown is the most dangerous kind of nuclear power accident. TMI was the most serious accident in the history of commercial nuclear power in the United States. Fortunately, not a single person from the nearby community or the plant itself was injured or killed as a result of the accident.

Environmental monitoring began within a few hours of the accident and continued long afterward. Careful estimates of radiation dose to the public determined that, on average, people living in surrounding communities received a radiation dose of much less than the equivalent of a chest x-ray. The worst-case dose for a hypothetical person standing at the site boundary would have been equal to the annual allowable radiation dose limits for members of the public (i.e., a small dose)—a fraction of the radiation dose that everyone receives annually from the natural environment. To date, health studies of the population in the surrounding area have not shown any adverse health impacts associated with the TMI accident.

Many communication and coordination issues in the response to this accident contributed to an atmosphere of uncertainty and mistrust. On the third day of the accident, following a controlled and intentional release of radioactive inert gas to relieve pressure in one of the vessels, the governor of Pennsylvania, in consultation with the NRC, decided to recommend evacuation of the most vulnerable segment of the population from the 5-mile radius of the plant. This group encompassed all pregnant women and all children 5 years old or younger—amounting to approximately 3,000–4,000 people. That announcement led to evacuation of nearly 200,000 people from the area.

This accident led to significant changes in the nuclear industry: upgraded and strengthened plant design and equipment requirements; broader and more robust regulatory oversight; increased involvement of local, state, and

Figure 4.6
The Three Mile Island nuclear power plant, approximately 10 miles southeast of Harrisburg, Pennsylvania. The two reactors are the two smaller structures. The TMI-2 reactor, where the accident occurred, is the one in the back. TMI-1, in the foreground, is still operating. The two large smokestacks were the cooling towers for TMI-2 and are no longer used. Not seen in the picture are two smokestacks associated with TMI-1. (Photo: U.S. Department of Energy.)

federal authorities; enhanced training and staffing requirements; and other improvements. On the negative side, it held back expanding use of nuclear technology for power generation in the United States for decades.

> *Did you know?* On March 22, 1975, a fire was started at Browns Ferry nuclear power plant in northern Alabama when a worker, who was checking for air leaks in a cable spreading room, used a lighted candle. The plastic foam used for sealing caught fire, and the fire spread and burned for several hours, damaging electrical cables and disabling a number of control systems. The reactor was shut down using alternative measures, thus averting a major radiation emergency. Major fire prevention and safety upgrades were instituted for all nuclear power plants as a result of this accident.

Case Study: Chernobyl

The April 26, 1986, Chernobyl accident in Ukraine, then part of the Soviet Union, is by far the worst nuclear power plant accident in the world. A series of poor operator decisions, together with a poor reactor design, resulted in a sudden power surge that ultimately caused a steam explosion, loss of containment, and a graphite fire that melted the reactor core and released 5% of that reactor core material into the environment over a period of 9 days. At the time of the accident, 600 workers were present on-site. Two workers died within hours of the explosion and 134 workers received high enough doses of radiation to develop radiation sickness. Of this group, 28 workers died within the first 4 months.

In 1986 and 1987, approximately 200,000 workers were involved in cleanup and recovery efforts. These people were referred to as "liquidators." The average radiation dose they received was equivalent to 10 whole-body CT exams. The first of the liquidators who reported to the accident on the first day received the highest doses; several of them received fatal doses (Figure 4.7). Subsequently, workers were rotated through the reactor for short periods of time and received lower doses of radiation. Eventually, liquidators increased to a number between 600,000 and 800,000, and most of these workers participated in off-site cleanup activities and received lower doses of radiation.

The radioactive materials released from Chernobyl spread over large areas of Belarus, Ukraine, Russia, Eastern Europe, Scandinavia, and, later, Western Europe. Radioactive material from the plant was detectable at very low levels over practically the entire Northern Hemisphere. However, most of the contamination was deposited on the surrounding area.

It took almost 1 week after the accident to evacuate residents of the contaminated town of Pripyat, where most of the plant workers lived (Figure 4.8). Next, the population of 116,000 people within the 30 km (18 mile) radius of Chernobyl was evacuated and they were eventually permanently relocated.

Figure 4.7

One of several monuments erected to honor the memory of "liquidators," who were the first responders to the accident. This monument was erected near Chernobyl. (Photo: Petr Pavlicek/IAEA.)

The exclusion zone was later extended to cover an area of 4,300 km^2 (1,660 mile^2). In subsequent years, under government orders, an additional 210,000 people in Ukraine, Belarus, and Russia were evacuated from their homes and resettled in less contaminated areas. The town of Pripyat, like a ghost town, remains inside the exclusion zone where only authorized entry is allowed. A new town, Slavutich, was constructed outside the exclusion zone, some 50 km (30 miles) away.

No one off-site of the Chernobyl plant developed radiation sickness. The most significant health impact for the surrounding population resulted from drinking milk contaminated with radioactive iodine, and this largely affected children and adolescents. Approximately 4,000 thyroid cancers have been detected in this group of children and 99% have been successfully treated. Unfortunately, this is a health impact that could have been averted if people had received timely recommendations and adequate alternative sources of milk (see Chapters 6 and 12). Iodine deficiency in this population may have also contributed to increased sensitivity to radioactive iodine and increased frequency of thyroid cancers.

Figure 4.8
The town of Pripyat, 3 km (2 miles) from Chernobyl, is now a completely abandoned city. The Chernobyl plant is seen in the background. (Photo: Petr Pavlicek/IAEA.)

On the positive side, to date there is no scientific evidence of any other significant radiation-related health effects to people exposed to radioactive material from Chernobyl. However, a slight increase in cancer incidence among the liquidators may be observed in the future.

The damaged reactor was one of four operating nuclear reactors at the Chernobyl site. The three remaining reactor units were restarted after the accident. One was shut down in 1991 after a serious fire in the turbine building. One was closed in 1996, and the last unit was closed in 1999.

Unlike the case with TMI, where the remaining nuclear materials from the damaged reactor core were removed from the damaged reactor unit and properly disposed,* the damaged Chernobyl reactor still contains a very large inventory of radioactive materials—nearly 95% of the original, but now melted, reactor core. In the months following the accident, a temporary concrete cover (or tomb) was hastily built to control further release of material from the reactor into the environment. This structure, referred to as a "sarcophagus" has since deteriorated (Figure 4.9). Through international

* This cleanup activity took 14 years to complete (August 1979 to December 1993) at a cost of $1 billion (www.threemileislandinfo.com).

Figure 4.9
The Chernobyl reactor with the sarcophagus cover. The construction of this cover was hastily completed in the months after the accident. The purpose of this deteriorating cover is to contain the large amounts of radioactivity inside the structure. The word "sarcophagus" means a stone coffin. (Photo: Petr Pavlicek/IAEA.)

collaboration, a new arch-shaped steel cover to slide over the deteriorating concrete cover is planned.

Did you know? The Chernobyl accident released approximately three quintillion Bq of radioactivity into the environment. A quintillion is equal to a million trillion, or 1,000,000,000,000,000,000 (1 + 18 zeros).

Did you know? According to the IAEA, the Chernobyl explosion put 400 times more radioactive material into the Earth's atmosphere than the atomic bomb dropped over Hiroshima. However, the combined atomic weapons testing conducted in the 1950s and 1960s is estimated to have put between 100 and 1,000 times more radioactive material into the atmosphere than the Chernobyl accident.

CRITICALITY ACCIDENTS

The types of nuclear reactions that occur in a nuclear power plant are called nuclear chain reactions. Such reactions are desirable in a reactor, but not in an industrial facility processing nuclear materials. It is possible that such a reaction accidentally takes place if mistakes are made in processing certain types of nuclear materials—usually some form of uranium or plutonium. In this case, intense gamma and neutron radiation is released and can inflict serious radiation injury and possibly death in people (usually workers) within a few meters of the accident. This is called a criticality accident. Even though the accident is serious and potentially lethal to those in close proximity, the reaction cannot be sustained due to physical and chemical limitations. Therefore, it is impossible for a nuclear explosion (nuclear bomb) to occur as a result of a criticality accident.

Several examples of criticality accidents have been documented around the world. One relatively recent example occurred in Tokai-Mura, Japan, in 1999. Three operators were engaged in processing a solution containing enriched uranium. Because of a deviation in approved procedures, a different type of container was used to handle the solutions and a criticality accident occurred. The two workers handling the solution received lethal doses of radiation. One worker died within 3 months of the accident and the second worker died within 7 months. Considering the high doses of radiation they received, the length of time they lived after the accident is unusual but can be attributed to the medical care they received, including bone marrow transplants.

Because the building was so heavily contaminated, high radiation levels could be measured outside. Approximately 160 residents of surrounding households within a 350-m radius of the building were evacuated, but they were allowed to return after 2 days. As a precaution, 300,000 residents within a 10-km radius of the facility were asked to remain indoors for 1 day, until the contaminated facility was properly shielded.

One more feature about criticality accidents is noteworthy. Because nuclear reactions occur in a criticality accident, neutrons are emitted and neutron radiation has the ability to make certain other materials radioactive. This process is called "activation." The ordinary material that has now been made radioactive is called "activation product." After a criticality accident, neutrons dissipate instantaneously, but the radiation coming off any activation products may still be measured by radiation detection instruments. For example, in the Tokai-Mura accident, radioactive sodium atoms could be measured in table salt taken from residents. Radioactive gold atoms could be measured in jewelry. This does not necessarily present a health concern, but it does demonstrate the high sensitivity of radiation detection equipment.

NATURAL DISASTERS

Most people may be surprised to hear that shortly after Hurricane Katrina struck New Orleans in August 2005 and rescue operations got underway, a U.S. Department of Energy aircraft was flying over the area looking for any signs of elevated radioactivity in the area. The concern was for radioactive sources owned by flooded hospitals or abandoned businesses because there was a potential for these sources to be unsecured. Also, a radioactive source manufacturing facility was located not far from the New Orleans area. Radiation control authorities in Louisiana and Mississippi worked to contact the owners and account for these sources.

It turned out that the radioactive source manufacturing facility was unaffected and secured and the aircraft flyover did not identify any radiation hazards in the flooded areas. Natural disasters such as wildfires or tornadoes can damage facilities and businesses containing radioactive material and they can potentially become lost or dispersed. Such a phenomenon has not yet occurred or been documented. Maintaining a current listing of all licensed radioactive sources and up-to-date contact information can help radiation control authorities identify and account for such sources if this situation should ever occur.

Resources

An excellent resource for more in-depth information about past radiation accidents is the IAEA Web site. IAEA publications can be found at http://www-pub.iaea.org/mtcd/publications/publications.asp

IAEA publications on accident response, including many accidents not covered in this chapter, can be found at http://www-pub.iaea.org/mtcd/publications/accres.asp

For more information on the Goiânia accident, see the following three sources:

IAEA. 1988. The radiological accident in Goiânia. Vienna, Austria (http://www-pub.iaea.org/MTCD/publications/PDF/Pub815_web.pdf).

IAEA. 1988. Dosimetric and medical aspects of the radiological accident in Goiânia. Vienna, Austria (http://www-pub.iaea.org/MTCD/publications/PDF/te_1009_prn.pdf).

Rozenthall, J. J., C. E. de Almeidat, and A. H. Mendonca. 1991. The radiological accident in Goiânia: The initial remedial actions. *Health Physics* 60 (1): 7–15. This was a special issue of *Health Physics* with articles covering various aspects of the Goiânia accident.

For general information about emergency planning for nuclear power reactors, see the Federal Emergency Management Agency (FEMA) Web site (http://www.fema.gov/hazard/nuclear/index.shtm).

For more information about the Three Mile Island accident, see the following three sources:

United States NRC. Fact sheet on the Three Mile Island accident (http://www.nrc.gov/reading-rm/doc-collections/fact-sheets/3mile-isle.html).

Report of the President's commission on the accident at Three Mile Island. October 1979. The need for change: The legacy of TMI. Washington, D.C.

Walker, S. J. 2004. *Three Mile Island: A nuclear crisis in historical perspective*. Berkeley: University of California Press.

For more information about the Chernobyl accident, see the IAEA Web site (www.iaea.org/NewsCenter/Focus/Chernobyl/#) or the following United Nations report:

United Nations Scientific Committee on the Effects of Atomic Radiation (UNSCEAR). 2000. Annex J: Exposures and effects of the Chernobyl accident. In *Sources and effects of ionizing radiation: Report to the General Assembly, with scientific annexes*, vol. II: Effects (http://www.unscear.org/docs/reports/annexj.pdf).

A comprehensive catalog of past radiation accidents is available from this Web site: Johnston, W. R. Database of radiological accidents and related events. Johnston's archive (http://www.johnstonsarchive.net/nuclear/radevents/index.html).

If you are interested in reading more about categorization of radioactive sources according to their danger levels, see IAEA. 2005. Categorization of radioactive sources. Safety guide no. RS-G-1.9. Vienna, Austria (http://www-pub.iaea.org/MTCD/publications/PDF/Pub1227_web.pdf).

5

INTENTIONAL RADIATION EMERGENCIES

OVERVIEW

UNFORTUNATELY, IT IS POSSIBLE FOR A small group of individuals intentionally to inflict great harm on a large population. Professionals working to prevent these attacks from occurring or preparing to respond to such attacks refer to these as CBRN (pronounced "see-burn") threats. CBRN is an acronym that stands for *c*hemical, *b*iological, *r*adiological, and *n*uclear. Another variation is CBRNE (pronounced "see-burn-ee"; the "E" stands for "explosives"). These acronyms, commonly used worldwide, refer to deliberate acts of terrorism aimed at inflicting great physical harm, panic, or societal disruption. You may see the acronym used in combinations such as CBRNE attacks, CBRNE events, CBRNE training courses, etc.

In this chapter, we describe radiological and nuclear threats. As with accidental radiation incidents described in the previous chapter, acts of terrorism involving radiation have a wide range of consequences, from minimal to catastrophic. You will learn a number of commonly used terms, such as improvised nuclear device (IND), radiological dispersal device (RDD), silent sources, dirty bombs, nuclear fallout, and radioactive plume, as well as several other concepts to understand the nature and impact of various types of radiation emergencies. In describing each scenario, we also explain a number of parameters that can directly affect how serious the consequences can be on the individual scale and for large populations.

There is no doubt that a nuclear detonation incident will inflict great harm on a large number of people and cause unthinkable destruction in a large metropolitan area. Other, non-nuclear radiation scenarios may result in a limited number of fatalities; contamination of buildings, city blocks, or agricultural areas; and a potentially significant economic loss. However, these radiological, non-nuclear scenarios have a true impact when they can create panic, uncertainty, and mass disruption. In this sense, the public can determine how such an incident can affect them by how they react and respond to it.

RADIOLOGICAL OR NUCLEAR SCENARIOS

People unfamiliar with emergency preparedness terminology may ask, "Is there a difference between radiological or nuclear emergencies?" or "Why do we need to separate the 'R' and 'N' categories when we consider CBRNE threats?" In the context of radiation emergency preparedness and response, as presented in this book and as used by various response agencies, the difference between a nuclear incident and a radiological incident is very simple. A nuclear incident involves detonation of a nuclear weapon (i.e., the proverbial mushroom cloud). Any and all other radiation incidents that do not involve a nuclear detonation are called radiological incidents.

The loss of life, loss of local response organization infrastructure, degree of physical devastation, and long-term consequences of a nuclear detonation are much more severe than in any other type of radiation incident. Therefore, for planning purposes, a distinct category is used to separate nuclear detonation incidents from radiological incidents. Whenever the general term "radiation emergencies" is used, it encompasses both nuclear and radiological scenarios.

The exception to this is emergency management professionals concerned with the safety of nuclear power plants. They have historically used the term "nuclear emergency" to refer to an emergency related to a nuclear power plant. This is not the terminology used in the broader context of emergency response planning. Any incident involving a nuclear power plant is considered to be a radiological incident because it cannot result in a nuclear detonation.

Following are examples and additional information on the use of the terms *nuclear* and *radiological* in other contexts (some readers may choose to skip ahead to the next section):

If you ask a number of scientists about the difference in meaning between the words "nuclear" and "radiological," you are likely to get many different answers, depending on which context the respondents use to frame their answers. Typically, the term "nuclear" refers to something originating from or relating to the nucleus. For example, gamma rays are "nuclear" radiation because they originate from the nucleus of an atom. However, x-rays do not originate from the nucleus and are not "nuclear" radiation, even though they may have the exact same properties as gamma rays.

In academia, an engineering school may have a nuclear and radiological engineering program. If engineers are concerned with optimizing the design and construction of nuclear fuel assemblies, they are engaged in nuclear engineering. If they are concerned with proper, safe handling of nuclear fuel assemblies and their subsequent disposal, they are engaged in radiological engineering.

In a 2002 IAEA publication, the term "nuclear or radiological emergency" is always used in combination—never separated—and repeated 81 times throughout the document. In its glossary, "nuclear or radiological emergency" is defined as "an emergency in which there is, or is perceived to be, a hazard due to: (a) the energy resulting from a nuclear chain reaction or from the decay of the products of a chain reaction; or (b) radiation exposure."* Many professionals may have difficulty interpreting this definition or applying it to the question asked in the beginning of this section: "Is there a difference between radiological and nuclear emergencies?"

As it pertains to the subject matter of this book, you can ignore all these nuances and definitions. You only need to remember that a nuclear incident occurs if, and only if, there is a nuclear detonation. Otherwise, a radiological incident has occurred.

IMPROVISED NUCLEAR DEVICE

An improvised nuclear device (IND) is a crude, low-yield atomic bomb or nuclear weapon. A nuclear weapon produces a nuclear detonation involving fission (or splitting) of atoms. As a result of these fission reactions, a tremendous amount of energy is released in a fraction of a second, creating an immense shockwave and intense heat. The release of energy in this form will cause great destruction and loss of life if the bomb is detonated near a populated urban area. Furthermore, the release of radioactivity and radiation as a result of this detonation can present a serious health hazard to people who survive the blast.

The bomb dropped over Hiroshima on August 6, 1945, code-named Little Boy, was by definition an IND (Figure 5.1). The science and technology of making nuclear weapons progressed rapidly over the next two decades and resulted in the development of weapons that were 1,000 times more powerful than Little Boy. These high-yield strategic nuclear weapons were a major concern during the Cold War, and they still require advanced state-supported technical sophistication to design, build, and maintain. However, a crude nuclear weapon similar to the bomb dropped over Hiroshima is relatively easy to engineer once the ingredient nuclear materials are at hand.

---

* International Atomic Energy Agency. 2002. Preparedness and response for a nuclear or radiological emergency, 58. Vienna: IAEA.

Figure 5.1

"Little Boy" is in the pit ready for loading into the bomb bay of *Enola Gay* at Tinian Island, where it took off for Hiroshima. Note the bomb bay door in the upper right-hand corner. The design of this bomb was simpler than that of the "Fat Man" plutonium bomb dropped over Nagasaki. (Photo: U.S. National Archives 77-BT-115.)

Therefore, a terrorist organization is more likely to obtain a crude or improvised nuclear weapon (i.e., a low-yield weapon).

In the following paragraphs, we introduce a few terms and concepts that are useful in understanding the impact and consequences of an IND detonation.

> *Did you know?* When Little Boy detonated, only 0.6 mg (0.02 ounces) of mass converted to energy. This is a manifestation of Albert Einstein's well-known scientific formula, $E = mc^2$, which describes how a tiny amount of mass can be converted to an enormous amount of energy. When used for peaceful purposes, there is virtually no limit to the potential that this scientific principle can offer.

Weapon "Yield"

The destructive impact of a nuclear weapon is directly related to its explosive yield, which is a measure of the amount of energy the weapon has released. The yield is expressed in terms of kiloton (KT) equivalent of trinitrotoluene, commonly known as TNT. If you consider the explosive energy that 1 ton of TNT would release, a kiloton is 1,000 times that amount, or 1,000 tons of TNT. Obviously, no amount of TNT is used in a nuclear weapon. However, its explosive yield is commonly expressed in equivalent terms that can be compared to the explosive yield of TNT.

An IND may yield an explosion in the 1- to 10-kiloton range; this is comparable to the bomb detonated over Hiroshima, which had an explosive yield of approximately 15 kilotons. For comparison, the conventional ammonium nitrate explosives used in the 1995 bombing of the Murrah Federal Building in Oklahoma City had an explosive yield of approximately 2 tons of TNT or 0.002 kiloton. With their sophisticated design and engineering, strategic nuclear weapons in the modern military can have explosive yields in the megaton range, or millions of tons of TNT.

Nuclear explosions release energy in three different forms: thermal energy (intense heat), shockwave (force of the blast itself), and radiation energy. As the yield of a nuclear weapon increases, the area of destruction it leaves behind becomes larger, especially as it relates to its thermal and blast effects. Contrary to what most people believe, the immediate and massive destruction by a nuclear bomb is *not* caused by radiation. We will discuss these three forms of energy release in the following sections.

> *Did you know?* If every person on the planet consumed a 2,000-calorie diet per day, the energy equivalent of that diet, for the entire planet, would be equal to a 13-kiloton nuclear explosion per day!*

Immediate Destruction (Heat and Blast Effects)

The heat and blast effects are the most destructive forces of a nuclear explosion. Nuclear reactions instantly generate a blinding flash of light with intense heat. Temperatures inside the fireball immediately after a nuclear explosion can reach tens of millions of degrees Fahrenheit—much higher than surface temperatures of the Sun. Approximately 35% of the total energy released from a nuclear weapon is in the form of intense heat that vaporizes most material in its immediate vicinity and causes severe burn injuries some distance away from ground zero.

---

* This assumes 6.6 billion people consuming 2,000 calories per day. There are 4.2 joules (J) in each calorie; 4.2 billion J is equivalent in energy to 1 ton of TNT.

Approximately 50% of the total energy released from a nuclear weapon is released as the force of the blast itself.* This shock wave, quickly following the immediate and intense heat, destroys buildings and throws large numbers of objects, including human bodies, broken glass, and debris, into the air at high speed, causing significant numbers of death and injuries. This can be particularly devastating in a populated urban area. The incendiary effect of this phenomenon is also significant because collapsed and damaged structures, leaking gas lines, and the spread of combustible material everywhere cause multiple fires, which in turn can result in additional damage and loss of life.

In the case of a 10-kiloton IND explosion, lethal shockwave injuries will extend to 600 m (approximately 0.4 mile) from ground zero in all directions. Lethal burn injuries will extend to 1,800 m (approximately 1 mile) away from ground zero in all directions. If a nuclear detonation were to occur in a modern urban environment, some of these effects would be different. People in basements have a better chance of surviving this initial impact and may be able to walk away if they are not trapped under fallen debris. The heat and blast effects of a nuclear weapon cause the greatest number of casualties, by far, than radiation alone (Figure 5.2).

Radiation from a Nuclear Blast

Only 15% of the total energy released from a nuclear weapon is in the form of radiation. However, due to the penetrating nature of the radiation produced and the way radiation interacts with matter, including living tissue, radiation and radioactivity from a nuclear detonation present serious health hazards to survivors. Survivors of the initial blast encounter radiation in two forms: initial prompt radiation from fission reactions at the time of detonation and a residual fallout radiation. Before discussing prompt and fallout radiation separately, let us briefly review the specific types of radiation and radioactive materials produced as a result of a nuclear explosion.

After a nuclear explosion, radiation and radioactive materials of any consequence can come from four sources:

- Neutrons and photons generated as a by-product of the nuclear chain reactions inside the bomb. These are produced at massive amounts

* This assumes a fission bomb (e.g., an IND) detonated on ground surface or up to an altitude of 30 km (100,000 ft). At higher altitudes, there is less air with which the energy of the exploding weapon can interact. As a result, the proportion of energy converted to shockwave and blast is reduced and more of the energy is converted to heat. (Glasstone, S., and P. J. Dolan. 1977. *The effects of nuclear weapons*, 3rd ed. Washington, D.C.: U.S. Department of Defense.)

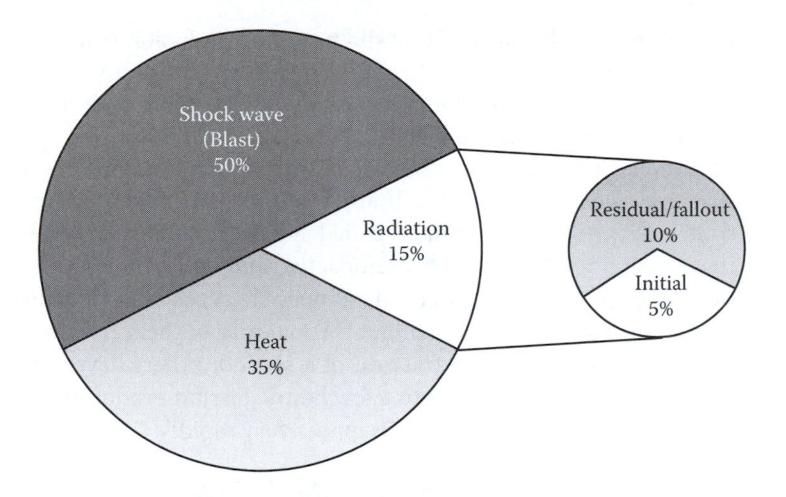

Figure 5.2
Nuclear explosions release energy in three different forms. Contrary to what most people believe, the immediate and massive destruction by a nuclear bomb is *not* caused by radiation. The heat and blast effects are the most destructive forces of a nuclear explosion. This chart shows the distribution of energy that would be released from an IND. (Adapted from Glasstone, S., and P. J. Dolan. 1977. *The effects of nuclear weapons*, 3rd ed. Washington, D.C.: United States Department of Defense.)

at the instant of detonation. These forms of radiation (neutrons and gamma rays) are penetrating radiation and can reach considerable distance from ground zero instantly.

- Activation products. Neutrons generated from nuclear chain reactions in the bomb will interact with otherwise ordinary materials (e.g., soil, building materials) and activate certain atoms in these materials, making them radioactive. Virtually any material that has trace amounts of iron, manganese, aluminum, silicon, sodium, chlorine, and cobalt, for example, can become activated by neutrons. These newly created radioactive atoms are called activation products. In the explosion zone, a great quantity of activation products will be sucked into the mushroom cloud. At distances away from ground zero, activation products will still be present, but in lesser quantities because the number of neutrons reaching those distant locations will be considerably fewer.
- Uranium and/or plutonium atoms that were part of the nuclear bomb construction, but did not undergo fission or nuclear reaction. After any nuclear detonation, there will be some amount of

this leftover fissile material that did not undergo fission reactions. Poorly constructed bombs "fizzle" and leave even more of this material behind. These unconsumed fissile materials emit alpha particles, beta particles, and gamma rays.

- Fission products. The fissile material (uranium and/or plutonium) that undergoes fission in the bomb nuclear chain reactions will split and form smaller atoms, called fission products. Many of these fission products are also radioactive and will emit beta particles and gamma rays. Hundreds of possible types of radioactive fission products are generated, but the majority of them have very fast half-lives, as fast as a fraction of a second; others have half-lives in the order of minutes to a few hours. Fission products with short half-lives will decay and disappear very rapidly.

Rule of thumb: Radiation levels 48 hours after detonation are 100 times lower compared to radiation levels at 1 hour after the blast.

Some fission products with longer half-lives (in the order of several years to decades) will be discussed later in this section under "Environmental Impact." Nearly all of the radioactive materials just described will get sucked and carried up into the resulting mushroom cloud.

Did you know? The bomb detonated over Hiroshima contained 64 kg (140 pounds) of uranium. However, only 0.7 kg (1.5 pounds) of that uranium underwent fission reactions, and only 0.6 mg (0.02 ounces) of mass was converted to energy.

Prompt Radiation Initial or prompt radiation is generally defined as all radiation emitted within the first minute after the nuclear blast from initial nuclear reactions and the radioactive material in the fireball and the radioactive cloud. The use of the 1-minute time window is somewhat arbitrary. One minute is used as the average time that it takes for the cloud of radioactive materials to rise and reach a high enough altitude so that only an insignificant amount of radiation will reach the ground. The radioactive materials in the cloud will then settle onto the ground and literally "fall out" of the sky to form a residual source of radiation that, by itself, can present a serious health hazard.

From a survival point of view, the complex radiation field at or near ground zero is irrelevant because it is highly unlikely that anyone will survive the extreme heat and blast effects. The neutrons and gamma rays that escape the explosion zone can instantly travel considerable distances away from ground zero. In the case of a 10-kilotom IND explosion, this initial prompt radiation can expose people to a lethal dose of radiation out to a radius of 1,200 m (approximately 0.75 mile) from ground zero in all directions.

Because of the geometry and physics involved, this radiation attenuates (loses its intensity) drastically with increasing distance from ground zero. If there are obstacles, such as walls or buildings, the radiation emitted from ground zero loses its intensity even faster and may get completely stopped by those obstacles. This is due to the principles of distance and shielding. (Refer to Chapters 3 and 9 for more information about these radiation protection principles and how they are applied in a nuclear emergency.) Therefore, the distance of a person from ground zero at the time of the blast and whether the person happens to be inside a building or in a basement will affect the amount of prompt radiation to which the person is exposed. For all practical purposes, exposure to prompt radiation is instantaneous.

> *Rule of thumb:* People far enough from the initial blast who are not physically injured by it and suffer no burn injuries are unlikely to have received a life-threatening dose from the prompt radiation and will most likely survive. Nuclear fallout would still present a significant danger to them in the next several hours and days after the blast. However, exposure to nuclear fallout can be considerably minimized or even avoided if appropriate protective actions, such as evacuation or seeking immediate shelter, are followed.

Nuclear Fallout

When a nuclear bomb is detonated at or near the ground surface (as is likely in a terrorist incident), everything inside the fireball vaporizes and large amounts of material, soil, water, asphalt, building material, and debris are kicked up, sucked into the cloud, and carried upward. Much of this material will be vaporized initially. Radioactive materials remaining or generated from the nuclear bomb will mix with the vaporized material in the mushroom cloud. This includes the radioactive fission products, radioactive activation products, and radioactive uranium and/or plutonium that were not consumed by the bomb's nuclear chain reaction.

As this radioactive cloud cools, it becomes condensed and forms relatively large dust particles. These condensed radioactive particles then begin to fall out of the sky and settle on the ground within minutes after the explosion. This is where the term "nuclear fallout" is derived. Deposition of fallout particles occurs mostly in an irregular downwind direction and can result in dangerously high and immediately life-threatening levels of radiation to a distance of several kilometers or miles downwind, depending on meteorological conditions.

> *Terminology:* The term "ground shine" is sometimes used to describe the radiation emitted from radioactive material deposited on the ground. Starting in the first hour and extending to several days after a nuclear detonation, ground shine can be a significant source of radiation, and people need to take precautions to minimize or avoid exposure to it.

Nuclear fallout is radioactive and will cause contamination of anything it lands on, including building and street surfaces, cars, rooftops, agricultural crops, animals, and people, if not properly sheltered. For people, prolonged exposure to fallout radiation can result in accumulation of enough dose of radiation to cause radiation sickness and death. In addition, nuclear fallout presents a long-term health and environmental hazard. Weather conditions play an important role here. Rainfall in the hours after the nuclear detonation can cause intense localized concentrations of nuclear fallout. Typically, raindrops containing fallout particles are dark in color; they have been referred to as "black rain."

On the other hand, wind, especially at higher altitudes, can carry the fallout particles relatively long distances away. How far away depends on environmental factors including the weather and terrain. With high-yield nuclear weapons, the mushroom cloud will reach into the upper layers of atmosphere and radioactive particles will be carried much farther distances.

It is important to recognize that radiation can easily be detected using commonly available instrumentation or by laboratory analysis. Therefore, radioactive fallout from an IND may be detected a couple of hundred miles away from the blast site; however, it is likely to be at very low concentrations. Thus, radiation levels will pose no immediate health hazard to people that far away, even though instruments can still measure it. In Chapter 2, we said that remnants of nuclear fallout from the atmospheric nuclear weapons testing done in the 1950s and 1960s can still be detected in your own backyard. Detecting radioactivity does not necessarily mean that it is doing any harm. Radioactively speaking, it is perfectly safe for children to play in your backyard.

We will describe in later chapters the importance of listening for and following instructions from local emergency management organizations after any radiation emergency. For example, evacuation (when not advised) may actually place people in a contaminated area and in harm's way.

Ground, Air, or Shallow Underwater Bursts

An air burst means that the nuclear weapon is detonated in air (e.g., dropped from an aircraft), so the fireball does not reach the ground surface. A ground burst means that the fireball reaches the ground surface, as is likely to be the case if terrorists use a truck to deliver the weapon. What makes the distinction between an air burst and ground burst important is that when detonation is at or near ground surface, large amounts of soil and building materials are vaporized and sucked into the cloud, followed by cooling, condensation, and formation of highly radioactive, relatively large dust particles, which will then settle on the ground.

In the case of an air burst where the fireball does not reach the ground surface, the resulting cloud of debris does not contain any vaporized material from the surface, although it still contains a large amount of radioactive material left over from the detonation (Figure 5.3). These radioactive materials will also condense and form particles. However, the amount of debris in the cloud is much less, and the resulting condensed particles are much smaller and will stay in the air longer before they settle on the ground. The lighter, short-lived radioactive particles have more time to decay in air before reaching the ground.

Figure 5.3

Mushroom cloud rising 60,000 feet over Nagasaki on August 9, 1945. Both Hiroshima and Nagasaki bombs were air bursts. They detonated between 500 and 600 m (1,600 and 2,000 ft) above ground. This was by design to increase the effects of the blast and, consequently, the amount of fallout was considerably less. (Photo: U.S. Department of Defense.)

Radioactive clouds reaching higher altitudes also will be carried farther away and radioactive particles will be more dispersed before they settle on the ground. As a result of all this, the amount of nuclear fallout settled near the site of detonation will be far less in an air-burst scenario than it would be in the case of a ground burst, and the fallout particles settled away from the site will have much less radioactivity.

Another potential scenario is a shallow underwater explosion. As border security measures are enhanced, it is more difficult for potential terrorists to bring a nuclear weapon into the country. To avoid risk of detection, such a weapon may be brought near a major harbor and exploded in shallow waters. This type of explosion has been studied before (e.g., the Baker test as part of Operation Crossroads in the Pacific) (Figure 5.4).

Figure 5.4

The Baker test detonated at Bikini Atoll on July 24, 1946, was a 23-kiloton shallow underwater detonation. The cauliflower-shaped cloud was condensation covering the water column inside. The water was highly radioactive and contaminated the ships nearby. The surge was 30 m (94 ft) at a distance of 300 m (1000 ft) from the center of explosion. (Photo: U.S. Department of Defense.)

The resulting water waves are a unique feature of this type of detonation. To use the Baker test as an example, the water wave was 30 m (94 ft) at a distance of 300 m (1000 ft) from the center of explosion. The extent of damage from such a detonation depends on many factors, such as the explosive yield, depth of water at the point of explosion (surface to sea bottom), depth of bomb at the time of explosion, and the distance to shore. In any case, the waves can cause devastating damage and the receding, highly contaminated seawater will also cause widespread high-level contamination and claim many lives.

Electromagnetic Pulse

Electromagnetic pulse (EMP) is an effect that is most severe and far reaching in cases of very high-altitude nuclear detonations. Even though an IND is not likely to be a high-altitude burst, some discussion of EMP may be helpful. The intense radiation field that is present immediately after a nuclear detonation will cause extensive ionizations in air, producing a massive number of electrons. As these free electrons move in the Earth's magnetic field, they create a massive electromagnetic current in the air. A high-altitude burst can affect communication and electrical systems 1,000 miles away.

However, in the case of a surface or low-altitude burst of an IND, the effect is much more limited. Most ionization events are contained within a few hundred yards of the blast and the electrical equipment at risk of being damaged by the electromagnetic pulse is within a small radius of about 5 km (3 miles) of ground zero. It is possible for local cell towers and communication systems that survive the initial heat and blast to be affected by EMP. This could negatively impact the ability of first responders to communicate with each other and for officials to communicate with the public. EMP does not harm people directly, but it may impair the function of cardiac pacemakers or other implanted electronic devices.

Health Impacts

The health effects of exposure to radiation are described in detail in Chapter 6. In this section, the immediate health impact of a nuclear explosion followed by the short-term and long-term consequences are briefly described in general terms. In Hiroshima, an estimated 60,000 people died immediately from the blast and heat effects of the bomb. Within a few months of the bombing, the number of deaths had reached an estimated 90,000–166,000 people, from a total population of 340,000–350,000. Among the approximately 90,000 survivors who have been studied in the decades following the bombing, only a few hundred cases of cancer have been attributed to radiation.

Figure 5.5

The devastated landscape of Nagasaki after the nuclear explosion and ensuing fires. The explosive force of the Nagasaki bomb was larger than Hiroshima's; however, the area of devastation was smaller in Nagasaki, partly because of the surrounding hills, which protected some areas against the blast. (Photo: U.S. Department of Defense.)

As we describe the health impacts of an IND in a major modern city, many factors can influence the number of casualties. Population density is an obvious factor that can contribute to an increased number of casualties. The types of building materials can make a difference as well. Modern building structures offer more protection than the typical housing structures in Hiroshima and Nagasaki in 1945 (Figure 5.5). Furthermore, the casualty rate can be influenced by the degree to which the local response infrastructure remains operational and the effectiveness of emergency response organizations addresses the needs of the injured.

Immediate Effects Within a small radius around the nuclear blast site, no humans will survive. Intense heat and effects of the shockwave will have claimed those lives instantly. Outside this radius, people who survive the immediate impact will suffer moderate to severe skin burns and/or blast-related injuries from flying debris or fallen structures. The extent of these injuries will depend on an individual's distance from ground zero and where he or she was at the time of the blast (e.g., inside a building, facing or away from windows, in a basement, etc.). Some of these injured survivors will receive high enough doses of radiation to develop radiation sickness. In medical terminology, this is referred to as "combined injury."

In addition, a nuclear blast is like an immensely bright flash of light. Because of this, people who may otherwise be a safe distance away, but happen to look directly in the direction of the blast, can suffer from serious eye injury. This can range from temporary flash blindness to severe burns on the retina and permanent blindness. The retinal burn injury can occur many kilometers away from the blast and is more severe if the blast occurs at night, when pupils are more dilated.

Short-Term Effects Survivors who suffer from combined injuries—blast injuries, thermal skin burns, and doses of radiation high enough to cause acute radiation syndrome or radiation sickness—will be at great risk. Many of these people may perish over a period of days or weeks, even with prompt medical care. However, this assumes that these initial survivors of the blast are rescued from the blast site and cared for in an expeditious manner before they succumb to their injuries.

Realistically, the secondary fires and high radiation levels will impede the search and rescue efforts; many people who survive the initial blast will die in the blast zone without being rescued. People who are far enough away and do not have any skin burns or blast-related injuries are more likely to have received lower doses of radiation because, as discussed earlier, radiation intensity drops rapidly as the distance from the blast site increases.

A serious short-term concern, however, is the impending nuclear fallout. Especially in the case of a ground burst, the radioactive fallout can result in localized high areas of contamination. Exposure to fallout radiation can be life threatening. Therefore, people should follow the instructions and protective action recommendations provided by emergency management authorities in order to minimize or avoid exposure to fallout. This topic will be discussed in detail in Chapter 9.

The displaced population could number in hundreds of thousands or even millions of people and would present a public health challenge because they would require temporary shelters and, in some cases, permanent shelters. In addition, many people would also require monitoring for radioactive contamination and assistance with decontamination and other health and medical needs.

Long-Term Effects The long-term health impact for many of the survivors may be some measurable, but relatively small increase in cancer rates among the population who were exposed to low or moderate doses of radiation. This is a population who would likely need long-term medical monitoring, possibly for several decades after the blast. Young children are especially at risk. Mental health issues cannot be overemphasized: Many people will lose family and friends and their livelihoods. The psychosocial impact of radiation emergencies will be discussed further in Chapter 8.

Environmental Impact

The residual radioactivity after a nuclear detonation makes the recovery from such overwhelming amounts of destruction even more challenging. As we described earlier, the radioactivity levels after a nuclear detonation drop rapidly in the hours and days following the detonation (Figure 5.6). In the first 2 weeks after a nuclear fission detonation, radiation dose rate decreases by a factor of 10 for each sevenfold increase in time. This means that the

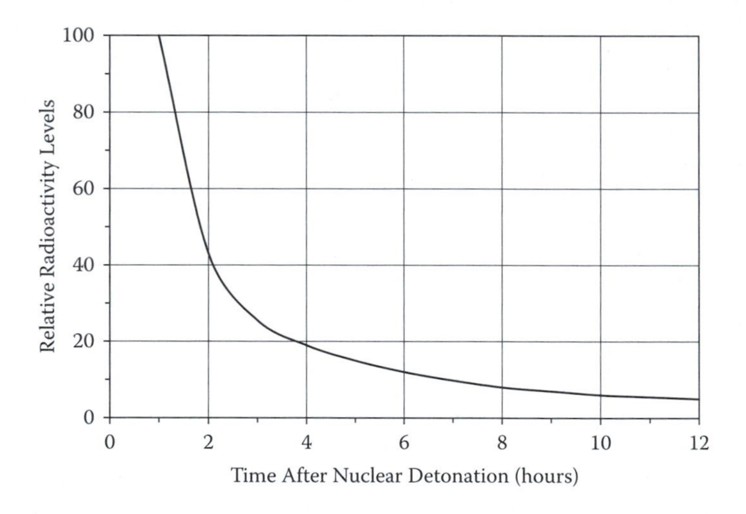

Figure 5.6
Radioactivity levels after a nuclear detonation drop rapidly because of the short half-lives of many fission products. Within 7 hours, the activity will have decreased to 10% of what it was at 1 hour after detonation. Within 2 days, the radioactivity will have decreased to 1%. See the footnote on page 91 for additional information. (From Glasstone, S., and P. J. Dolan. 1977. *The effects of nuclear weapons,* 3rd ed. Washington, D.C.: United States Department of Defense.)

radiation dose rate 2 days after detonation is only 1% of the dose rate at 1 hour after detonation. At the end of 2 weeks, radiation levels are down to 0.1% (1/1000) of what they were an hour after detonation.*

A number of fission products have longer half-lives. Most notably, cesium-137 (Cs-137) has a 30-year half-life. Strontium-90 (Sr-90) has a half-life of 29 years. Therefore, some of the radioactivity from the nuclear bomb will linger in the environment for many years, even decades later. A devastated urban area can be cleaned and rebuilt, however. The best living example is the city of Hiroshima; for some years now, it has been a beautiful, thriving metropolitan area inhabited by millions of people.

> *Did you know?* A radioactive isotope of iodine, I-131, which is widely used today in nuclear medicine applications, is a fission product. Following a nuclear detonation, I-131 is dispersed into the environment, but it has a short half-life of 8 days.

A Common Misconception

Some people may mistakenly believe that an accident in a nuclear power reactor could result in a nuclear detonation similar to bombs detonated over Hiroshima and Nagasaki. The process of nuclear fission and the need to sustain a nuclear chain reaction are basic principles of operation in nuclear power generation. However, the physical, chemical, and engineering preconditions required to bring about a nuclear detonation do not exist in a nuclear power plant. The worst-case scenario at a nuclear power plant would be a meltdown of the reactor core, a breach in the reactor containment walls, and release of large amounts of radioactivity into the surrounding environment—a scenario similar to what occurred in the Chernobyl accident in 1986 (Chapter 4). This would be catastrophic, indeed, but a nuclear detonation cannot occur even if the power reactor is heavily damaged or sabotaged.

RADIOLOGICAL DISPERSAL DEVICES

A radiological dispersal device (RDD) is a tool specifically designed to spread radioactive material and contaminate humans or the environment.

* This relationship is meant as an approximation. It is based on the assumption that the fallout at the location descends in a very short period of time. In reality, fallout may continue to descend for several hours after detonation. Therefore, this relationship is not an exact predictor of what the radiation levels will be, but it does demonstrate how rapidly radiation levels decrease in the hours and days immediately following a nuclear detonation.

The method of dispersal can be by any number of means (e.g., explosion, spraying a liquid, aerosol). The most talked about example of an RDD is the so-called "dirty" bomb. In constructing a dirty bomb, perpetrators can use any conventional explosives (such as dynamite, C-4 plastic explosive, or a homemade fertilizer bomb) and package it with any radioactive material they can find. When the explosives are set off, in addition to the destructive effect of the explosion, the blast disperses the radioactive material into the surrounding area (Figure 5.7).

The most important point to keep in mind is that, in the case of an RDD, it is the conventional explosion that causes destruction and fatalities, if any. The radioactive component of a dirty bomb is aimed at creating panic by contaminating people and the places where they work, live, or engage in commerce. In other words, the purpose is to create societal disruption—not destruction. For this reason, RDDs are sometimes referred to as "weapons of mass disruption."

The radioactive material in a dirty bomb will no doubt complicate the rescue and recovery operations, but it will not prevent them. It will also complicate long-term recovery and cause a significant financial burden. What is most important, however, is to communicate effectively with the public and all stakeholders and to provide them with timely and accurate information.

After an RDD explosion, radioactivity levels are expected to be high in localized areas very close to the blast site. First responders, *following proper safety procedures,* can perform rescue operations and assist the injured who may be contaminated. We discuss these topics later in Chapters 11 and 14.

Some of the radioactive material from the blast will be dispersed in the surrounding area. One factor affecting the dispersal is the chemical form of the radioactive material—fine powder, metallic pellets, liquid, or ceramic. Another factor is the device geometry—how the RDD has been constructed. If the radioactive material is in certain chemical forms and the RDD has been constructed well to allow the radioactive material to aerosolize as fine particles, some of the radioactive material will be carried away with the speed and direction of the wind. This "plume" of radioactive particles may spread over a large area, but the more it spreads, the less concentrated it gets and the less of a health concern it becomes.

Just as in the case with nuclear fallout long distances away, the plume of radioactive material from a dirty bomb may be carried a considerable distance downwind (e.g., tens of kilometers) and be detectable with common radiation detection instruments. However, the levels of radioactivity will be low. We discuss in Chapter 9 what steps people in surrounding areas can take to protect themselves, but it should be noted that radiation levels will be far lower than what would be necessary to cause immediate health concerns. It is important for emergency management and public health authorities to

Figure 5.7

The site of a dirty bomb incident could look like this scene. This is the Murrah Federal Building in Oklahoma City after the April 19, 1995, terrorist bombing that claimed 168 lives and injured hundreds more. If the truck bomb had contained some amount of radioactive materials, the number of casualties would not have changed, but the damaged building, debris, and surrounding areas would have become contaminated, further complicating the recovery effort. (Photo: U.S. Department of Defense.)

provide timely and accurate information and instructions to the public, and it is important for the public to follow the instructions.

In summary, some key points regarding an explosive RDD or a dirty bomb include:

- A dirty bomb is not a nuclear bomb and does not cause a nuclear explosion.
- How the material is dispersed depends on the physical and chemical form of the radioactive material, how it is physically packaged with the explosives, and how much explosive was used.
- Better dispersal of radioactive materials actually lowers the concentration and reduces the radiation hazard—likely down to insignificant levels.
- Radiation can be easily measured at very low concentrations and any potential radiation hazards can be quickly identified by trained professionals and proper instrumentation.

Shrapnel

In some cases, the radioactive material from an RDD may be dispersed in the form of small fragments. If this occurs, these fragments scattered in close vicinity of the blast will result in areas of high radiation levels, which could present a serious radiation hazard. First responders will have to monitor the site for such areas. It is unlikely, but not impossible, for such a fragment to be embedded in the body of an injured or deceased victim. If so, the presence of such fragments can be confirmed quite easily using a portable radiation detector. Fragments can then be removed and properly disposed.

Nonexplosive RDDs

An RDD may use nonexplosive means to disperse radioactive materials. For example, radioactive material in water-soluble form can be sprayed using a tanker truck, a small airplane (e.g., crop duster), or even manually by a hand-sprayer. Radioactive materials can also be spread in the form of a powder or aerosol, contaminating the air supply in a populated building. On the one hand, nonexplosive RDDs do not cause the immediate injury or fatality of a dirty bomb. On the other hand, because a nonexplosive RDD occurs without a "bang," it can go undetected or unreported for some time, perhaps weeks. As a result, radioactive contamination will be spread to many people and many places and may even cross international borders via unsuspecting travelers.

The levels of radioactivity in these cross-contaminating events will likely be miniscule with no potential health risks. However, for a number of people who may come in direct contact with the radioactive material at the site of dispersion, the levels of radioactivity can be very high. For these people, internal contamination with radioactive materials will be highly likely because they may eat with their contaminated hands or inhale contaminated particles. For these individuals, radiation sickness and possibly death are potential outcomes if their symptoms go undiagnosed for days and weeks. This topic is further discussed in Chapters 6 and 11.

Case Study: Goiânia

In Chapter 4, the 1987 radiological accident in Goiânia, Brazil, was described in detail. In many respects, that accident resembles the case of a nonexplosive RDD. The radioactive material was unknowingly dispersed by a Brazilian family, contaminating themselves, neighbors, and relatives. The accident went unnoticed for 3 weeks. During that time, many individuals suffered from internal contamination and, ultimately, four died in spite of receiving medical care. Another important circumstance was the large number of people (112,000) who requested to be monitored and screened out of concern that they might have been contaminated. This is likely to be the case with almost any intentional RDD incident.

Case Study: Poisoning in London

In November 2006, the poisoning and subsequent death of a Russian dissident, Victor Litvinenko, made international headlines. Allegedly, he was poisoned with the radioactive isotope polonium-210, and a few days later he reported to a London hospital as he began to fall ill. He died 19 days after admission to the hospital due to multiple organ failure. It was only 2 days before his death (i.e., 17 days after he was first admitted to the hospital), that radioactive poisoning was even suspected to be the cause of his illness.

This incident, even though it apparently targeted only one individual, had similarities with a nonexplosive RDD: Radioactivity was not suspected for many days and many public places around London, even a football stadium seat, were found to be contaminated with trace amounts of radioactive polonium. These were places that either Litvinenko or the crime perpetrator had apparently visited. The majority of these contaminated locations did not present a public health concern, according to British public health authorities. The most contaminated locations were a restaurant, a hotel bar, a hotel room, and Litvinenko's apartment (Figure 5.8).

Figure 5.8
This restaurant in downtown London was one of several locations where polonium-210 contamination was found. While the restaurant remained closed for cleaning the radioactive polonium, an iconic James Bond gun barrel sign covered its doors with a note about the "international espionage incident." The restaurant has since reopened. (Photo courtesy of Dr. Robert Whitcomb.)

The British authorities worked tirelessly to reach members of the public who might have visited these contaminated locations and offer them testing. This outreach effort involved, in addition to British citizens, people in 52 different countries around the globe and the public health agencies in those countries, including the United States.

Health Impact

The health impact of an RDD incident would be orders of magnitude less than that of an IND detonation. An explosive RDD will damage property, may cause a limited number of casualties, and will spread radioactive contamination in the surrounding areas. However, only a few people, if any, may develop radiation sickness. A nonexplosive RDD may also result in a number of deaths in a worst-case scenario and is likely to contaminate a large number of people to varying degrees, some with high levels of internal contamination.

A potentially major health impact of an RDD incident is the likelihood that large numbers of uninjured people may report to hospitals out of concern or suspicion that they may be contaminated or may need medical assistance. This will present a serious challenge to hospital emergency rooms that are already overburdened under normal day-to-day operations. The public health community is fully aware of this potential, and many emergency response plans include public communication strategies to help minimize this impact.

Screening the potentially affected population for external and internal contamination will be a major public health challenge as well. How efficiently this screening is performed depends on many factors, including the type of radioactive material used (alpha emitter or gamma emitter) and how prepared the local and state response organizations are.

Environmental Impact

A radiological dispersal device, by definition, will contaminate the environment. Even if people are the target of contamination (e.g., the poisoning incident in London), the environment will also become contaminated as people move around and interact with that environment, especially public places. Therefore, the extent of environmental contamination and its consequences can vary widely under different incident scenarios. Contamination may be spotty and only in a number of discrete localized areas. Contamination may cover large areas of farmland or it may cover several city blocks. As the environmental impact of radioactive contamination is evaluated, these questions should always be asked:

- What specific radionuclide is the contaminant?
- Exactly where or what is contaminated?
- How much radioactivity is present?
- What is the potential for human exposure?

The answers to these questions will help determine any risks to human health and how extensive cleanup and recovery efforts might be. Environmental effects of radiation emergencies are described in detail in Chapter 7.

Economic Impact

The economic impact of an RDD incident can be high. These weapons are intended to be weapons of mass disruption. Their aim is to disrupt society and commerce. An RDD incident could contaminate New York City's financial district, requiring potentially costly and lengthy cleanup. Simultaneous RDD incidents in a few major cities at the height of the holiday shopping season would have significant economic consequences. Dollar estimates are occasionally made to associate specific monetary values to economic losses that might result from such hypothetical scenarios. The estimates are invariably high. Economic consequences of an RDD incident are influenced by two factors:

- Physical destruction caused by the incident. This is rather self-explanatory and includes damage to buildings and critical infrastructure (bridges, tunnels, etc.).
- Public reaction to the incident. How the public responds to such incidents may have an even greater impact on the economy than the effect of any physical destruction. Engaging in commerce is, of course, vital to a healthy economy. Effective public communication and taking steps to ensure the public safety can help restore people's confidence, but this is certainly easier said than done. If a metropolitan area's downtown is contaminated with radioactive materials, how long will it take to clean the contamination? How clean is "clean enough"? How much will it cost? How long before workers are willing to return to critical infrastructures, such as banks, that may have some small but fixed residual contamination? These decisions will need to be made with input from all stakeholders. The same principles will apply to food and agricultural products from areas that may have been contaminated. Applying objectivity will lessen the economic impact of all these decisions.

RADIATION EXPOSURE DEVICE

A piece of radioactive material may be intentionally placed in a public place, hidden from sight, with the sole purpose of exposing people to radiation without their knowledge. Such a device is called a radiation exposure device (RED). It may also be called a radiation emitting device, a hidden source, or a silent source; however, all this terminology refers to the same type of incident. An RED may be placed in public transportation (under a bus or subway seat), a busy shopping mall (the food court, for example), movie theater, or any other location where a large number of people may sit, stand, or pass close by.

The radioactive source, used in this manner, is likely to be a "sealed" source. This means that the radioactive material is encased in some form of plastic or metal housing and does not leak. Therefore, radioactive contamination is not an issue. Only penetrating radiation leaves the source container and can expose or irradiate people nearby. Individuals who come in contact with or even touch or sit on a sealed source do not become contaminated (see Chapter 3 for the distinction between contamination and exposure). The danger is from exposure to high levels of radiation in close proximity to the source for extended periods of time, assuming that the RED is a radiation source emitting high levels of radiation.

If the radioactive source is not sealed, or if the seal around the source is somehow breached, some of the radioactive material could be released and cause contamination. At that point, the device is more like a radiation dispersal device and people coming in contact with the source could spread contamination elsewhere.

Intentional use of a sealed radioactive source by a Chinese national to target and irradiate a business associate was reported in the Chinese media in 2002, but details of that incident were unclear. That incident was allegedly an attempted murder targeting one individual, rather than a terrorist incident.

Health Impact

The health impact of an RED will depend on the characteristics of the radioactive source, the specific radionuclide, and the amount of radioactivity. It will also depend on how close people get to the source and how long they stay close to the source. The next chapter will show that the physical symptoms of radiation sickness may take some time (days or weeks) to develop. Even then, it may take more time to associate those symptoms with radiation exposure and subsequently isolate where the source may be located. Therefore, many people could be exposed to radiation before the RED is discovered. Of course, not everyone irradiated by the source will develop radiation sickness.

A characteristic of an RED is that it behaves like a small localized source, referred to as a "point source." As distance to a point source increases, radiation levels drop sharply (see Figure 9.1). Therefore, many people coming into the vicinity of the radioactive source would only receive a small dose of radiation, if any. These people would certainly not experience any immediate health effect, and it is unlikely that they would suffer any long-term health effects either. In the example of an RED taped to the bottom of a bus seat or a subway seat, passengers sitting three or four seats away from the source fall into this category; that is, they would likely receive a small dose of radiation and adverse health effects would be unlikely.

On the other hand, an individual sitting on the seat with the radioactive source under would most likely develop symptoms if the period of exposure was long enough. How long is long enough depends on how "hot" the radioactive source is. The expression "hot" refers to the amount of radioactive material or radiation levels. For heavily exposed individuals, symptoms of radiation sickness are likely to develop and death could possibly follow.

What is helpful in this situation is that only a portion of the body would experience the high radiation exposure (in this case, the buttocks), rather than the whole body. Therefore, there is a higher tolerance for the dose, even though that part of the body would be very critically injured.

In summary, an RED can cause severe radiation injury, even death, for a small number of people. A larger number of people will be exposed to small, most likely inconsequential doses of radiation. Just like the case with RDDs, REDs are intended to create fear. A declaration that several REDs had been placed in shopping malls across the country could create a great deal of anxiety. Making that announcement on Black Friday (the day after Thanksgiving in the United States) or the Saturday before Christmas, which are both busy shopping days, could create sensational results for terrorists.

Environmental Impact

As described earlier, an RED is a sealed source: The radioactive material is housed in a container and does not leak. Once the radioactive source is located, it can be safely removed. No amount of contamination is left behind in that location. In short, once the radioactive source is found and removed and the lack of contamination verified by sampling, the environmental impact of an RED is zero.

OTHER SCENARIOS

Contaminated food or water can be a method to inflict harm on one or more individuals. A well-known example is the 2006 radiation poisoning incident in London where radioactive polonium was served in a pot of tea. However, intentional contamination of food or water supplies with radioactive materials is not a plausible method to harm a *large* number of people. Even if a large amount of radioactivity is released in a community's water supply, it is significantly diluted to lower concentrations because of the large volume of water.

There is also the possibility that a radioactive source would be used in combination with a chemical or biological agent. This is not a likely scenario. However, if it did occur, health and safety considerations regarding the chemical or the biological agent would almost certainly take priority over concerns for the radiological agent.

Resources

Two books provide an excellent account from the perspective of several atomic bomb survivors from Hiroshima and Nagasaki. Both books are relatively small volumes; they are not technical, but rather descriptive of life in the first few moments and days following a nuclear detonation:

Hersey, J. 1989. *Hiroshima*. New York: Vintage Books. This book was originally published in 1946. This newer edition includes one additional chapter with an update on the lives of the individuals the author had chronicled four decades earlier.

Trumbull, R. 1957. *Nine who survived Hiroshima and Nagasaki: Personal experiences of nine men who lived through both atomic bombings*. New York: E. P. Dutton & Co. This is a true story of nine people who were originally from Nagasaki, but were working in Hiroshima on August 6, 1945. They survived the atomic attack on Hiroshima and managed to make their way back home to Nagasaki—just in time for the August 9 bombing. They survived that too.

A classic technical textbook on the description of nuclear explosions and its aftermath has been written by Glasstone, S., and P. J. Dolan. 1977. *The effects of nuclear weapons,* 3rd ed. Washington, D.C.: U.S. Department of Defense.

The following document provides a brief and informative description of an urban nuclear detonation incident and makes recommendations to local and state planners on how to prepare and respond to such an incident:

Homeland Security Council. 2009. Planning guidance for response to a nuclear detonation: First edition. Available from www.hsdl.org

For a summary of unclassified technical data from Sandia National Laboratory regarding the science of RDD explosions, particle characteristics, and hazard analysis, see Harper, F. T., S. V. Musolino, and W. B. Wente. 2007. Realistic radiological dispersal device hazard boundaries and ramifications for early consequence management decision. *Health Physics* 93 (1): 1–16.

6

HEALTH EFFECTS OF RADIATION

OVERVIEW

Chapters 4 and 5 depicted a variety of scenarios where exposure to radiation and contamination with radioactive materials can occur and briefly discussed the potential health impacts. This chapter describes the health effects of radiation in more detail. The health effects of exposure to radiation have been studied extensively for decades and are well known. The health impact of radiation exposure can range from mild effects, such as reddening of skin, to severe effects, such as death in the short term or cancer in the long term. What determines the potential outcome is first and foremost the dose, as is the case with exposure to any other harmful agent. In the case of radiation, dose means the amount of radiation energy delivered to the body (see Chapter 3). Multiple parameters, such as the type of radiation, the dose of radiation, method of exposure, and duration of exposure, can define the relative hazard of radiation exposure and influence its health outcome.

- If the radiation dose is *high* and is delivered to the whole body in a short amount of time, death is the likely outcome. Death can occur within a few days to several weeks after a period of severe illness, including internal bleeding and infections. If the dose is high enough, loss of consciousness can occur within a few minutes and death follows shortly thereafter.
- If the radiation dose is *moderate*, symptoms of radiation sickness will develop in the hours and days following the exposure in a rather predictable manner. The exposed individual has a good chance to survive. Chances of survival are even better for individuals who receive prompt medical care and have not suffered any burns or traumatic injuries. The individuals who survive this moderate dose of radiation will, however, have a higher risk of developing cancer later in life compared to the average person.

- If the radiation dose is *low*, there will be no radiation sickness or immediate health effects. There will likely be no observable health effects later in life either. Individuals exposed to low doses of radiation may have a slightly increased risk of cancer compared to the average population. However, when the dose is very low, this increased risk is too small to be measured or estimated and, for all practical purposes, is nonexistent.

We start this chapter by describing the basics of how radiation can damage the human body and continue with how various parameters, such as type of radiation, the rate at which the dose is delivered, partial body irradiation, and other factors can affect its health outcome. Understanding the difference that each of these factors makes is essential in putting any radiation exposure in its proper perspective. We end the chapter by describing the health effects using an easy-to-read question and answer format. Depending on their level of interest, readers may choose to skip ahead to this section to find quick answers to commonly asked questions, such as:

- What is the clinical presentation of radiation sickness and how much time after exposure will symptoms appear?
- What determines how bad or severe the symptoms will be?
- How much radiation exposure is fatal?
- Can radiation cause cancer?
- Is there a cure for radiation sickness?
- Are there any drugs or antidotes to radiation poisoning?
- Is anything not known about the health effects of radiation?

The references at the end of the chapter provide more in-depth discussion of these topics.

> *Terminology:* The terms "radiation sickness" and "radiation poisoning" are common phrases that are used interchangeably. A more clinical term to describe the same phenomenon is "acute radiation syndrome" (ARS).

RADIATION ACTION ON LIVING CELLS

This section gives a brief description of how radiation causes injury. Ionizing radiation is a form of energy and gets its name, "ionizing," from its ability to ionize atoms or molecules as it delivers its energy to the material through which it passes. Ionization means ripping one or more electrons from an atom. These atoms or molecules with unpaired electrons are called "free radicals" and are chemically very reactive.

Did you know? The most common pathway for radiation damage in human tissue starts with ionization of the most abundant molecule in the body: water. This interaction forms HO• molecules (called hydroxyl free radicals), a highly reactive chemical species.

Free radicals interact with other atoms or molecules to "steal" or share an electron. They undergo this reaction instantaneously (within fractions of a second) and, in the process, alter whatever with which molecule they interact. The process is called "oxidation." The molecule thus altered is said to have sustained oxidative damage. These damaged molecules could be lipid (fat) molecules that are part of the cell membrane, proteins inside the cell, or the DNA molecule itself. DNA is made of nucleotides (bases, sugars, and phosphates) and packaged by proteins. Oxidative damage to DNA could lead to degradation or loss of a nucleotide base, a break in the sugar-phosphate strand, or a cross-link or chemical joining with one of the surrounding protein molecules.

Once an important component of the cell like the DNA is damaged, three potential outcomes are possible (Figure 6.1):

- Cellular repair: As life has evolved, mechanisms to repair such damaged molecules have also evolved. Cells have a remarkable ability to repair DNA damage or simply to remove the damaged piece and replace it with a new part. Such repair and maintenance activities

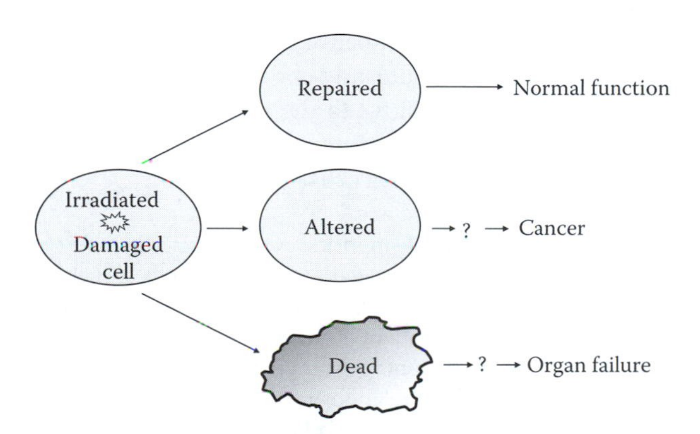

Figure 6.1

Potential outcomes after radiation causes damage to a cell. The cell may repair the damage and resume normal function. Misrepaired damage can create an altered or mutated cell, which may serve as a precursor to cancer. The cell may die and, if a large number of cells die, organ failure can result.

are part of a cell's everyday routine. When damages are successfully repaired or if the damage is to an unimportant, dispensable molecule, there will be no health consequences.

- Cellular alteration or mutation: Some molecular damages, however, are harder to repair than others and the cell can occasionally make an error in making such repair. This error can result in an altered base sequence, a missing piece of DNA, or a rearranged piece of DNA. If this altered or misrepaired part of DNA is a segment of a gene that is not expressed or used in that cell's daily function, there will be no consequence. However, if the change in sequence alters the expression of a critical gene or cell function, the cell is said to be mutated. Although a mutation can be the precursor of cancer, not all mutations lead to cancer.
- Cellular death: Another likely outcome for a radiation-damaged cell is death. This could result from cellular damage that is not repaired and interferes with a critical function such as cell division. It could also be a result of a damaged cell "committing suicide." The technical term for this is apoptosis (pronounced a-pop-toe-sis) and it is sometimes referred to as "programmed cell death." Apoptosis describes a series of genetically predetermined biochemical processes leading to cell death. In this context, apoptosis is believed to be an evolutionary mechanism for an organism's survival at the expense of a defective cell. However, if the radiation dose to an organ is high and a large number of cells die, that organ's function is impaired, resulting in acute (and chronic) health problems. Some body tissues are more sensitive to the effects of radiation than are others. In general, rapidly dividing cells are more sensitive to radiation damage than nondividing cells.

Did you know? Normal metabolism inside every cell produces free radicals and oxidative damages to DNA on a daily basis; thousands of such damages are produced every hour in every cell. Nearly all of these are processed and repaired. However, some are not, and the accumulation of unrepaired oxidative damages is believed to contribute to detrimental effects such as cancer and aging. Although these relationships are still not well understood, this is the basis for popularity of food and nutritional supplements that are rich in antioxidants. The oxidative damage produced by ionizing radiation has remarkable similarities to the types of DNA damage caused by normal metabolic processes. However, some evidence suggests that at least a fraction of radiation-induced damages is more complex and hence not as readily repaired as oxidative damages induced by normal daily metabolism are.

Did you know? Radiation burn is not really a "burn" in the traditional thermal sense. Typical skin burns as a result of fire or extreme temperatures

Table 6.1 Key Factors Affecting the Health Outcome of Radiation Exposure

Type of radiation
Radiation dose
Whole body or partial body irradiation
Acute or chronic exposure
External or internal exposure
General status of individual's health
Presence or absence of traumatic injury (burns or wounds)
Availability of medical care

are seen immediately. A radiation burn is caused by death of skin cells responsible for skin regeneration caused by radiation damage as described before. Although some early symptoms (such as itching and reddening) may appear quickly, it takes several days to weeks for a radiation burn to manifest itself fully; the outer skin layer, already dead, is shed, exposing deeper, now unprotected layers of skin tissue unable to regenerate a protective cover.*

KEY FACTORS AFFECTING HEALTH OUTCOME

In this section, several key parameters of importance that can define the radiation hazard or influence its health effects are described (Table 6.1). Readers may skip this section, depending on their level of interest, and go directly to the questions and answers section later in this chapter. However, understanding and appreciating the key parameters described in this section are necessary for evaluating any cases of radiation exposure and for putting them in proper perspective. The following material builds on the basic information provided earlier in Chapter 3. The reader may refer to that chapter, if necessary, for a quick review of some of the basic concepts used here.

Type of Radiation

Penetrating radiation such as photons (x-rays, gamma rays) and neutrons present an external radiation hazard. That means that from a source outside the body, they can penetrate and deliver their energy inside the body. Photons are a potential source for internal exposures as well if gamma-emitting radionuclides (such as cesium-137, iridium-192, or cobalt-60) are

* Skin graft is a possible treatment in severe cases.

accidentally ingested or inhaled. There are no plausible ways for x-rays or neutrons to generate from inside the body.

Alpha particles, on the other hand, are not penetrating radiation and therefore they are not an external hazard. Even if lethal amounts of alpha particles are just inches away, as long as they remain outside the body, they present no hazard as external radiation. Even directly placed or deposited on the skin, alpha contamination presents no health hazard; it is impossible for alpha particles to penetrate the outer dead layer of skin and damage the live cells underneath (Chapter 3). If the contamination makes its way inside the body, however, the internal exposure by alpha particles can have serious health consequences.

Alpha particles deposit large amounts of energy over short ranges. The high density of ionizations damages the cells the alpha particles go through and can affect tissue and organ function in the body. Inhaled alpha-emitting radioactive materials can deliver large doses of ionizing radiation to the lung tissue. In the case of americium-241 (an alpha emitter), the material tends to be absorbed and taken up by the bone, resulting in a disproportionately large dose to the bone compared to other body organs.

Beta particles are more penetrating than alpha particles and can penetrate the outer layer of skin and damage the living cells underneath, potentially causing a severe radiation "burn." But if beta particles are just a few feet away, they present no external hazard. If beta-emitting radioactive materials are inhaled or ingested, they present an internal hazard.

The chemical and physical natures of the radioactive material are also important determinants in cases of internal exposure. How soluble the material is and how quickly it is excreted from the body will influence the radiation dose delivered to the body and, consequently, the health effects.

> *Did you know?* When high-energy beta particles are emitted inside the body (e.g., from strontium-90 atoms present in trace amounts in all bodies), the beta particles slow down in body tissues. As they lose energy during deceleration, they emit photons that are called bremsstrahlung radiation, a German term meaning "braking radiation" or "decelerating radiation." The photons from bremsstrahlung radiation are technically defined as x-rays. In that sense, x-rays are generated from inside the body! These levels of bremsstrahlung x-ray radiation, however, are far too low to be of any health consequence.

Radiation Dose

The single most important parameter in determining the adverse health effects from radiation exposure is the magnitude of the dose. In the very low range of radiation doses (<0.05 Gy or 5 rad), even at exposures several times the annual radiation dose from natural background radiation, one should reasonably expect no adverse health effects. Certainly, no immediate health effects are observed as a result of these low doses, although the risk of getting cancer later in life may be slightly increased. In the midrange of radiation doses (1–4 Gy or 100–400 rad), symptoms of radiation sickness will manifest; chances of survival are good, especially with prompt medical treatment, but the risk of getting cancer later in life is increased. In the high-range of radiation doses (>8 Gy or 800 rad), symptoms of radiation sickness will manifest more severely and death is the likely outcome even with aggressive therapy (Table 6.2).

Table 6.2 Mortality after Acute Whole-Body Exposure to Ionizing Radiation

Radiation Dose[a]	Time to Vomiting[b] (% of Patients Vomiting)	Chance of Dying from Radiation Exposure		Chance of Dying from Fatal Cancer[c]
<0.05 Gy	No vomiting	0%	No clinical indications	20%
<1 Gy	No vomiting	0%	Mild temporary decrease in white blood cells	<25%
<2 Gy	>2 hours (up to 40%)	0%	Chance of survival improves with availability of medical care, the underlying health status of the individual, and the absence of combined traumatic injury such as burns or wounds	<30%
2–4 Gy	1–2 hours (up to 80%)	<50%		30–40%
5–8 Gy	<1 hour (80–100%)	>50%		45–60%
>8 Gy	<10 minutes (100%)	100%	Survival is unlikely even with aggressive therapy	n/a

Sources: Adapted from the National Academy of Sciences. 2006. *Health risks from exposure to low levels of ionizing radiation*, BEIR VII report. Washington, D.C.: National Academies Press, and Flynn, D. F., and R. E. Goans. 2006. *Surgical Clinics of North America* 86:601–636.

[a] 1 Gy = 100 rad.

[b] The time it takes to start vomiting after radiation exposure is a useful indicator for radiation dose when no other dosimetry or clinical information is available. The times listed are only approximate. Not all patients in a dose category vomit according to this schedule. Expected percentage of patients vomiting is given in parentheses.

[c] The baseline risk of fatal cancer in the absence of any radiation exposure is 20%. Any cancer resulting from radiation exposure will occur years after the exposure.

Threshold Dose A minimum dose is required to induce certain health effects of radiation exposure. That minimum dose is referred to as the threshold dose. Below the threshold dose, the effect is typically not observed. For skin reddening, for example, a minimum dose of 2 Gy (200 rad) to the skin is required. Below that dose, no skin reddening occurs. For temporary sterility in men, a minimum dose of 0.15 Gy (15 rad) to the testes is required. Below that dose, no temporary sterility will occur. For permanent sterility to occur in men, a dose higher than 3 Gy (300 rad) to the testes is needed. Permanent sterility in women requires a dose higher than 2 Gy (200 rad) to the ovaries. Temporary hair loss (men or women) results following exposure to a threshold dose of 3–5 Gy (300–500 rad). This means exposure to low doses of radiation does not cause permanent sterility or hair loss.

> *Terminology:* Health effects mentioned in the previous paragraph (skin reddening, hair loss, sterility) as well as vomiting, diarrhea, and cataracts are referred to as *deterministic effects* in the scientific literature. For deterministic effects to occur, a person must be exposed to a threshold dose of radiation. Beyond that, the higher the dose gets, the more severe the health effects will be. In contrast, there are also *stochastic effects*—health effects dependent on chance. The single most important stochastic effect of exposure to radiation is cancer. No minimum dose is required. Even a small dose of radiation is presumed to have some chance (albeit, very small) to cause cancer. It is the probability or the chance of getting cancer that is dependent on radiation dose—the higher the dose is, the greater the chance of developing cancer is. Once cancer forms, its severity is not dependent on the radiation dose that caused it. A higher dose of radiation does not result in a more aggressive form of cancer.

Whole-Body versus Partial-Body Irradiation

When a radiation dose is delivered to only one part of the body, it causes localized radiation injury, and much larger doses can be tolerated. For example, if a very high radiation dose of 30 Gy (3,000 rad) is delivered to a limb, the arm or leg will suffer serious injury, including severe skin burns; in a worst-case scenario, it may have to be amputated. The individual, however, will survive. But if the same dose is delivered to the entire body, death is the certain outcome.

Acute versus Chronic Exposure

This distinction refers to the period of time during which the radiation dose is delivered to the body. The exposure is defined as acute if the totality of the dose is given over a period of a few hours or less. If the dose is delivered over a period of a few days or longer, the exposure is defined as chronic. For example, your exposure to natural background radiation (Chapter 2) is a chronic exposure. But when you get a chest x-ray, that exposure is acute. For the same amount of total radiation dose, the more protracted the period of exposure (i.e., the longer it takes to deliver that dose), the less its health impact will be.

For example, in a lifetime of 80 years, a person is exposed to 0.25 Sv (25 rem) of radiation just from natural background radiation.* Certainly, this chronic exposure presents no health concerns. However, an acute exposure of 0.25 Sv (25 rem) will carry some measurable health risks and should be avoided if possible.

Two plus Two Does Not Equal Four A related concept is called dose fractionation, or the split-dose effect. When it comes to radiation doses, two plus two does not equal four. If a radiation dose is divided into two or more fractions and is delivered over an interval of time, the accumulated effect is less than if the total dose is delivered at once. The principle of dose fractionation is applied by doctors who use radiation to treat cancer. The most efficient way to kill tumor cells is to deliver the entire dose in one treatment. However, many normal, healthy cells and possibly the patient could die using such a treatment regimen. For this reason, the dose is fractionated or split and given over a period of many days. This allows healthy cells to recover between treatments.

Acute exposures are more of a concern than chronic exposures because the body has little or no chance to repair or overcome the radiation damage from acute exposure. For example, a person receiving an acute dose of 3 Gy (300 rad) to the whole body will exhibit signs and symptoms of radiation sickness and will have a 50% chance of death within 60 days if untreated. By contrast, the same dose given to the same individual in multiple fractions over a period of approximately 1 year is a survivable exposure, although it carries with it an increased risk of developing cancer later in life.

Dose Rate One of the most important pieces of information, especially for first responders, is the concept of dose rate—the amount of radiation dose delivered per unit of time. This piece of information will determine how long a person can safely work in a contaminated or high-radiation area before having to leave or turn back. For example, if the radiation level in a contaminated area is one roentgen per hour (R/h) and live victims are in need of rescue in that area, the first responders have 25 hours to work there before

* This total does not take into account radiation doses from medical exposures, which, in the lifetime of an average person, may have a magnitude similar to that of total natural background radiation.

they reach the recommended 250 mSv (25 rem) dose guidance for life-saving activities. If the radiation level is 50 times higher (i.e., 50 R/h), the first responders have only 30 minutes to work in that area before reaching the same dose guidance (more discussion on this topic is given in Chapter 14.)

External versus Internal Exposure

External exposure occurs whenever the source of radiation or the radioactive material is and remains outside the body. Radiation exposure (or irradiation) continues as long as the source of radiation is present. The exposure (or irradiation) stops immediately once the source of radiation is removed or shielded or as soon as the individual leaves the radiation area. Getting a chest x-ray or a CT scan is an example of external exposure: The irradiation stops when the machine is turned off. If radiation levels are high, they can cause serious injury. It should be noted that personal protective equipment, even the level A "moon suit," does not protect against penetrating external radiation such as gamma rays.

Internal exposure occurs when radioactive materials have somehow entered the body. This takes place primarily through inhaling air contaminated with radioactive particles, by ingestion of contaminated food or water, or by eating or smoking with contaminated hands. In the case of internal exposure, the radiation dose is delivered to the body from within. It is not as easy to remove the radioactive material from inside the body and you certainly cannot walk away from it. Therefore, radiation exposure continues as long as the radioactive materials remain inside the body. Some radioactive materials are completely eliminated from the body in a few days, while others can firmly deposit themselves in body tissues and remain in residual amounts in the body for life. Therefore, the period of irradiation can range from a few days to a few decades.

Some medical interventions can significantly accelerate excretion or removal of radioactive materials from the body (see Chapters 11 and 12). The adverse health effects of internal exposure depend on the radioactive materials themselves, the amount of radioactive material internalized, and whether the radioactive materials distribute uniformly in the body or become concentrated in a particular organ (e.g., bone, kidney, or thyroid). Most cases of internal exposure do not result in serious health consequences. However, under certain circumstances, internal exposure can lead to death (see the Goiânia accident and London poisoning case studies in Chapters 4 and 5, respectively).

Terminology: Some types of radioactive material tend to accumulate in a particular organ of the body. In such a case, that organ is said to be the "target organ" or "critical organ" for that particular type of radioactive material. For example, the target organ for iodine is the thyroid gland, the target organ for uranium is the kidney, and the target organ for radium, americium, and strontium is bone. Other radionuclides, such as cesium, distribute uniformly in the body in soft tissues, and the target organ is said to be the whole body. In cases where a specific tissue is a target organ, health effects that result from internalization are driven by the radiation dose to that specific organ or body tissue.

QUESTIONS AND ANSWERS

What Is Radiation Sickness?

Acute radiation syndrome (ARS), commonly known as radiation sickness or radiation poisoning, is a broad term describing a range of symptoms that develop when the body is exposed to a high dose of radiation in a short period of time.

What Are Symptoms of Radiation Sickness and How Long after Exposure Will Symptoms Appear?

In mild cases, nausea, weakness, and loss of appetite appear within a few hours and then resolve shortly afterward. If radiation doses are higher, nausea and vomiting occur within an hour or two, are more intense, and will persist for up to 2 days. The patient may develop a fever. The chance of opportunistic infection increases as the number of circulating white blood cells decreases. Internal bleeding may occur as the number of circulating platelets decreases. In more severe cases (higher radiation doses), gastrointestinal cramping and bleeding will develop within a few days.

In what is called a "latent period," symptoms usually appear to improve for a period of a few days or weeks (depending on dose). The latent period is much shorter with higher doses. Hair loss occurs mostly during this period. After the latent period, symptoms return with more severity. Patients again will experience loss of appetite, nausea, vomiting, bleeding, infections, bloody diarrhea, and dehydration. The cause of death in most cases is infection and internal bleeding. In the most severe cases, diarrhea develops immediately; the exposed individual becomes disoriented and may experience seizures or go into a coma shortly after radiation exposure. These latter signs and symptoms are associated with fatal outcomes.

Did you know? Nausea and vomiting, the early signs of radiation sickness, are also common symptoms of stress, fear, and anxiety. In an emergency, public concern over radiation exposure may lead individuals to experience nausea and vomiting. Health care providers (and others) may mistakenly regard these findings as an indication of radiation sickness.

What Determines the Severity of Symptoms?

Many factors can modify symptom severity in radiation sickness or dictate whether any symptoms will appear at all after radiation exposure. The two most important factors are the size of the radiation dose and the extent of the exposure (total or partial body). Whether an individual has preexisting medical conditions or coexistent trauma such as burns or wounds are also key factors. Traumatic injuries have a strongly negative impact on the likelihood of surviving a radiation injury.

Terminology: The term "combined injury" is used to describe a situation where a person is exposed to a high dose of radiation and is also suffering from burns, fractures, bleeding wounds, or other traumatic injury.

What Is a Radiation Burn?

Radiation burn refers to radiation damage to the skin not caused by thermal energy (heat). This damage can begin as a redness of skin (similar to a sunburn) within a few hours after exposure. The skin may become swollen and itchy. These effects will resolve after a short time. But if the radiation dose is high, they will return. Skin damage will worsen over the next few days to several weeks because of radiation-induced necrosis (death) of the cells responsible for skin regeneration. A deep wound with exposed flesh that is highly susceptible to infections is the result. Complete healing of the skin may take weeks to years, depending on the radiation dose. Although skin grafts are a possible treatment in some severe cases, they do not always take. In other cases involving extremities (arms and legs), amputation may become necessary.

Did you know? The Japanese survivors of Hiroshima and Nagasaki suffered severe burns from the heat of the nuclear explosions (Figure 6.2). These are different from radiation burns. Due to lack of proper medical care and medical supplies, cooking oil or even machine oil was used to "treat" some of these burn victims.

How Much Radiation Exposure Is Fatal?

If someone is exposed to a whole-body dose greater than 8 Gy (800 rad), chances of survival are small, even with aggressive medical treatment including bone marrow transplant. For doses less than 8 Gy (800 rad), chances of survival are dependent on the medical care that can be provided promptly,

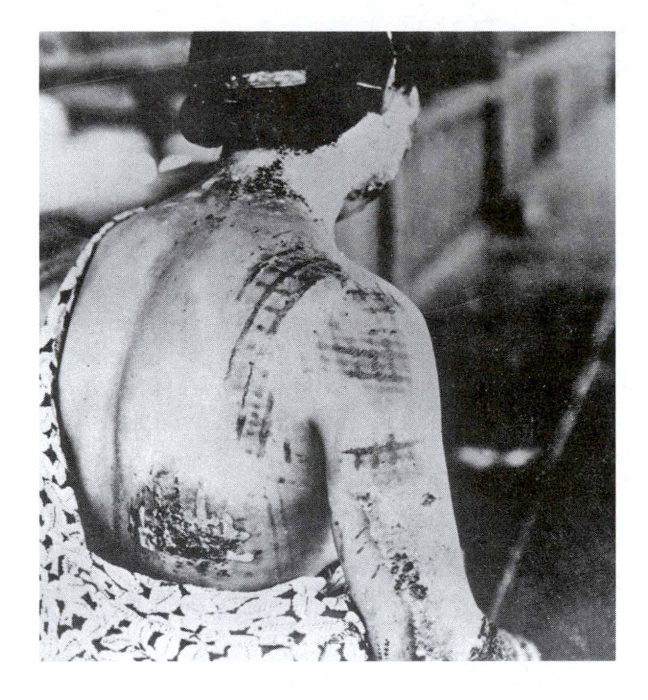

Figure 6.2
The burn pattern on the body of this Hiroshima victim corresponds to the dark portions of the kimono she wore at the time of explosion; dark colors absorbed more heat. This is not a radiation burn. (Photo: U.S. National Archives 77-MDH-65.)

the underlying health status of the individual, and the absence or presence of acute traumatic injury. For doses less than 1 Gy (100 rad), chance of survival is nearly 100%.

Terminology: A technical term used in toxicology to describe lethality of an agent is the mean lethal dose (LD_{50}), which means the dose that will kill 50% of those exposed. In the case of humans exposed to acute doses of radiation, the LD_{50} falls somewhere between 3 and 4 Gy (300 and 400 rad). With prompt and aggressive medical treatment, LD_{50} can be increased and more people will survive higher radiation doses. However, this level of medical care may not be available to all those exposed in a mass casualty situation (see Chapter 11).

Did you know? Although dogs are similar to humans in their sensitivity to radiation, rats can tolerate twice the radiation dose that people can. Cockroaches can tolerate and survive radiation doses 10-fold higher. A strain of bacteria called *Deinococcus radiodurans* can tolerate radiation doses several thousand times greater than the median lethal dose in humans.

Is Radiation Sickness Contagious?

No, radiation sickness is not comparable to sickness from bacterial or viral infections. Acute radiation syndrome is not caused by an agent that can be transmitted—no matter how closely a person comes in contact with, touches, sleeps next to, or interacts with others. It may be helpful to think of people who receive medical x-rays or CT exams and then go home. They have been irradiated, but they no longer carry radiation with them and cannot pass or transmit it to others. People exposed to high doses of radiation are similar. It is impossible for their vomit or any other body fluids to cause radiation sickness in others.

One note of caution: If radiation sickness is caused by high levels of internal contamination, some radioactive material will be present in urine, sweat, vomit, and stool. Using standard precautions (e.g., gloves, gowns, and face masks) will provide adequate protection for caregivers and can keep the radioactive material from contaminating them.* Even if standard precautions are not followed or if protective measures fail, radiation sickness is not something that can be transmitted. The worst outcome in this case would be contamination of a caregiver with radioactive materials. The amount of contamination, however, would be far short of the amount necessary to cause radiation sickness.

Can Radiation Sickness Be Cured?

Certain medical treatments can help patients suffering from radiation sickness and significantly increase their chances of survival. These include antibiotics to treat infections, fluids to keep patients hydrated, and medication to help the body's cellular immune system recover faster (Figure 6.3). Bone marrow transplants can also be used to treat victims of high-dose exposures when medications alone are not effective. See Chapters 11 and 12 for more discussion of these topics. If the radiation dose is extremely high—exceeding 10 Gy (1,000 rad)—chances of survival decrease sharply even when the best medical care is available. During triage in a mass-casualty situation, when

* The term "standard precautions" is an extension of "universal precautions" and is based on the principle that all blood, body fluids, secretions, excretions except sweat, nonintact skin, and mucous membranes may contain transmissible infectious agents. (From Centers for Disease Control and Prevention. 2007. Guideline for isolation precautions: Preventing transmission of infectious agents in healthcare settings. Available from www.cdc.gov).

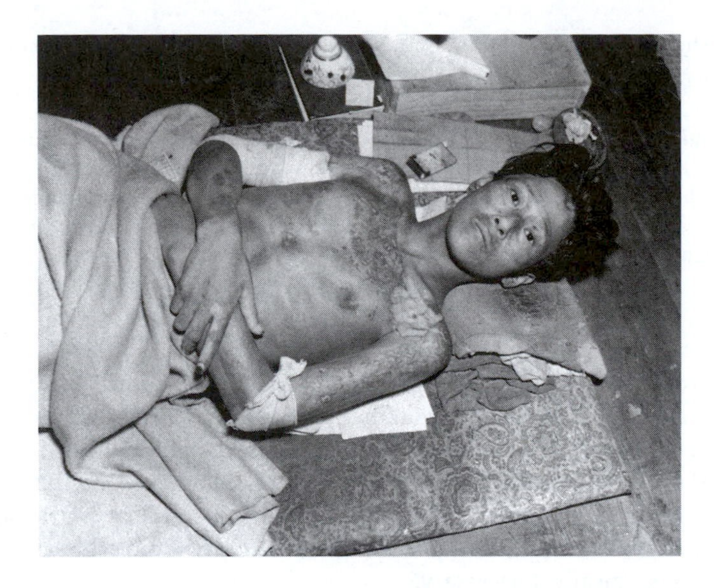

Figure 6.3
A Nagasaki victim lying in bed. Many victims with suppressed immune systems and burn injuries stayed in unsanitary conditions, which increased the risk of infection. (Photo: U.S. National Archives 77-AEC-52-4457.)

medical resources are limited, patients with little or no chance of survival will not be given priority for medical treatment.

What Is Radiation Poisoning and Is There an Antidote for It?

The term "radiation poisoning" is sometimes used synonymously with radiation sickness or acute radiation syndrome. Gamma rays and x-rays are not poisons that one can eat or drink; however, radioactively contaminated material can be ingested or inhaled. In such cases, some medications and treatment methods may be available to help the body get rid of contaminants faster. See Chapter 12 for a detailed discussion of this topic. It is important to note that these treatments are not antidotes per se and cannot immediately rid the body of the contamination or reverse any damage already done.

Can Radiation Cause Cataracts?

Yes, but this requires a minimum threshold dose of 2 Gy (200 rad) to the eye and it may take a few years after the exposure for the cataracts to develop.

How short this latency period is and how severe the cataracts will be depend on the radiation dose to the eye.

> *Did you know?* Radiation-induced cataracts are similar to those caused by aging, resulting in partial opacification (cloudiness) of the eye's lens. A layer of epithelial cells covers the lens of the eye. Radiation damage to these epithelial cells interferes with their proper function and prevents light from traveling straight through.

Are Adverse Health Effects Caused by Radiation Reversible?

Most radiation health effects can be reversed. The human body has a remarkable ability to recover from many radiation-induced injuries as long as the radiation dose is not extremely high. For example, hair loss is reversible. Many cases of skin damage are reversible. Temporary sterility, as the name implies, is reversible. Injuries to the body's blood-forming tissue, the immune system, and the intestinal tract are reversible. Even cataracts can be corrected surgically if necessary.

However, when the radiation dose is high enough, sterility will be permanent, and the damage to skin, bone marrow, lining of the intestine, cardiovascular system, and brain will become irreversible. The dose required to make such injury irreversible is different for each tissue. For example, it would take a smaller dose to cause permanent sterility than it would take to cause permanent brain damage and death. Patients exposed to a very high dose of radiation will die even with the best medical care available. Such care will only be successful in prolonging the inevitable, and during this time the quality of life will be sharply diminished. Fortunately, exposure to such high doses of radiation is very rare.

How Long Does It Take to Recover from Radiation Sickness?

Recovery depends on severity of the symptoms, but full recovery could take up to 2 years.

What about Cancer?

Ionizing radiation is a carcinogen. The probability (chance) that a given radiation exposure will cause cancer increases with the dose. Although cancer is one of the most commonly feared outcomes of radiation exposure, radiation is generally considered to be a weak carcinogen compared to baseline cancer risk. In the United States, for example, approximately 42% of the population develops some form of cancer in their lifetime due to natural causes;

approximately 20% develop fatal cancers. In other words, the odds of a U.S. resident developing some form of cancer over his or her lifetime due to natural causes are approximately 1 in 2.4. A radiation dose of 0.1 Sv (10 rem) increases the odds slightly to 1 in 2.3, and a dose of 0.25 Sv (25 rem) increases these odds to 1 in 2.2.

What do these numbers mean? If a large population is exposed to a low dose of radiation, a number of additional cases of cancer in that large population will be found that would not otherwise have occurred. But for any one individual exposed to the same dose of radiation, the odds of developing cancer are not significantly increased. This example may illustrate the difference: Imagine in a metropolitan area, the city announces that to help pay for the mayor's new leather chair, the police will issue five extra traffic tickets this year to five randomly selected motorists. The number of traffic tickets issued this year will increase by five, but for any given driver in this large city, the odds of getting a traffic ticket are not significantly increased.

> *Did you know?* A large medical follow-up study of Hiroshima and Nagasaki survivors has been in progress for the past 60 years. This is a population of 120,000 survivors, including a group of unexposed Japanese as the control group. At this point, approximately 20,000 cases of cancer have been found in this population, but fewer than 1,000 cases have been attributed to radiation exposure from the atomic bombs.

How Long Does It Take for the Cancer to Appear?

The time period between radiation exposure and appearance of cancer is called the latent period. Leukemia has the shortest latent period and can appear in as few as 5 years after exposure. Other types of cancer have significantly longer latent periods and can take decades to develop.

Does Radiation Cause a Different Type of Cancer?

No. Once any form of cancer is diagnosed, it is not possible to tell whether it was caused by a past radiation exposure or by natural causes. One exception is a form of leukemia called chronic lymphocytic leukemia (CLL). There is currently no link between CLL and radiation exposure and cases of CLL are not considered to be radiogenic (caused by radiation).

Are Children or the Elderly at Increased Risk of Injury from Radiation?

Yes, but the risks for each are different. Children under the age of 18 are more sensitive to the carcinogenic (cancer-causing) effects of radiation. This age dependency has been shown in a study of Japanese atomic bomb survivors. In the case of breast cancer, for example, young girls and teenagers are more sensitive to radiation than are adult women. Also, convincing data for thyroid cancer indicate that a young population is more sensitive to radiation than older adults are.

Another factor is also in play. When radiation exposure occurs at a young age, a considerably long period of time is ahead during which cancer can develop and be diagnosed. When an older population is exposed to radiation, their remaining life spans may be shorter than the latency period required for cancer to develop.

The older population, however, will be more sensitive to acute effects from large doses of radiation. It is more difficult for them to recover from radiation sickness because, to begin with, their general health status is poorer as a group compared to younger, healthier adults. In any radiation emergency, children and the elderly will be considered special populations and their needs will receive appropriate priority.

Can Radiation Damage Be Passed on to the Next Generation?

This phenomenon is called the hereditary effect of radiation. Radiation can damage germ cells—that is, cells that later turn into mature sperm for men or ova (eggs) for women. Suppose the radiation damage does not destroy the cell but rather causes a genetic mutation (a rare effect of radiation). Now suppose that such a genetically altered sperm or egg forms an embryo that survives to go through the gestation (pregnancy) and develops into an infant with a genetic disorder. This has been shown to occur in animal studies and, in theory, it can also happen in humans, although no cases have been documented. The children of the Japanese atomic-bomb survivors have been studied extensively for various indicators of birth defects; among that group, no radiation-related hereditary effect has been shown to exist. Only the children born to mothers who were already pregnant when the atomic bombs exploded showed increased incidence of microcephaly (small head size), growth retardation, and mental retardation.

When patients undergo a medical procedure that requires gonadal exposure to a large dose of radiation (such as radiation therapy for cancer), they may be advised to wait for some time before attempting a pregnancy. Any person who has had radiation therapy or a chemotherapy procedure should seek the advice of a physician about observing such a waiting period.

What If a Pregnant Woman Is Irradiated?

In the case of accidental radiation exposure or a radiation emergency, the good news is that the fetus is somewhat protected in the uterus and the radiation dose to the fetus from external radiation sources will likely be lower than it is to the mother. However, because the human embryo and fetus contain many dividing cells, they are sensitive to radiation, so severe health consequences, including growth retardation, malformations, mental retardation, and childhood cancer, can occur. Important determinants of effect include gestational age (weeks into the pregnancy) and radiation dose.

A fetus may also be exposed to radiation when a pregnant woman inadvertently ingests or inhales radioactive materials. Water-soluble isotopes may be absorbed into the maternal bloodstream and pass into the fetal circulation. Internal contamination may also concentrate in the mother's bladder, in close proximity to the womb. Again, possible health consequences depend on the age of gestation and level of contamination. A small amount of contamination is unlikely to cause significant maternal or fetal harm. Nevertheless, pregnant women are always a priority for receiving medical assessment in a radiation accident or emergency.

What about routine medical exposures? If you are pregnant, it is important to inform your health care providers so that any medical radiation exposures to your abdomen can be avoided or minimized. However, many diagnostic procedures do not involve exposing the abdomen to radiation or the radiation dose to the abdomen will be very small. An unfortunate reality is that in cases when such diagnostic procedures have been performed on pregnant women, it is not uncommon that therapeutic abortions may be considered as an option. Such abortions may be completely unnecessary. Women who have been exposed to radiation while pregnant or think they have been exposed to radiation should seek expert advice before considering termination of the pregnancy. For example, whereas a fetal dose of 0.1 Gy (10 rad) does have a small individual risk of radiation-induced cancer, there is a 99% chance that the exposed fetus will not develop childhood cancer or leukemia.

What about occupational exposures? If your profession involves handling radioactive materials or exposure to radiation, it is important to let your supervisor know, in writing, that you are pregnant so that alternate duty can be arranged.

> *Terminology:* The exposure of a fetus to radiation is referred to as prenatal radiation exposure. Any health consequences of radiation on the developing fetus are referred to as teratogenic effects. Of course, this is not unique to radiation. Drugs (both licit and illicit) and alcohol can produce teratogenic effects.
>
> *Did you know?* Women with otherwise uneventful pregnancies (and in the absence of any radiation exposure above natural background) have a 15% chance (or higher) of miscarriage and a 3% chance of delivering an infant with birth defects. Any added risk from radiation exposure should be evaluated considering this baseline.

Is Anything Not Known about Health Effects of Radiation?

The health effects of radiation have been studied for over a century. The one question often debated among radiation professionals is whether a very low dose of radiation—on the order of a few times the annual dose from natural background radiation—causes any increased risk for cancer. Radiation protection policies and practices in most countries are based on the assumption that an increased risk of cancer exists even at low doses of radiation. But this increase has not been demonstrated and is not likely to be shown experimentally because it requires a large irradiated population to study. The fact remains that even if low doses of radiation carry some increased risk of developing cancer later in life, this increased risk is very small.

Currently, active areas of research are investigating a number of interesting, but not well understood, phenomena occurring at very low doses of radiation. These include:

- Adaptive response: A small dose of radiation, under certain circumstances, has been shown to induce a transient (temporary) resistance to subsequent doses of radiation. This phenomenon has also been observed in cases of exposure to chemicals.
- Bystander effect: Irradiated cells grown in culture (i.e., in a laboratory) have been shown to interact with neighboring nonirradiated cells, and this interaction can have detrimental, and sometimes beneficial, effects on the nonirradiated cells. The molecular mechanisms of this action are still not well documented, and the effect is yet to be shown in organisms.
- Genomic instability: This is a phenomenon seen when cells recovering from radiation damage divide and produce new cells, which show a higher probability of undergoing mutations or chromosomal changes.

These research areas may ultimately lead to revision of current estimates of cancer risk from low radiation dose and dose rates, either upward or downward. But the estimates are not likely to change by much. Although radiation is a carcinogen, it is a weak carcinogen.

Resources

An excellent online resource for information on health effects of radiation is the Radiation Event Medical Management (REMM) Web site maintained by the National Library of Medicine (www.remm.nlm.gov).

For graphic images of radiation injury from past accidental exposures, see the International Atomic Energy Agency (IAEA) reports referenced in Chapter 4, visit the REMM Web site, or see the joint IAEA/World Health Organization brochure, "How to Recognize and Initially Respond to an Accidental Radiation Injury" (available from www.who.int/ionizing_radiation/a_e/IAEA-WHO-Leaflet-Eng%20blue.pdf).

For a thorough medical description of acute radiation syndrome and its separate components, the hematopoietic, gastrointestinal, cardiovascular, and central nervous system syndrome, see:

Mettler, F. A., and A. C. Upton. 2008. *Medical effects of ionizing radiation,* 3rd ed. Philadelphia: W. B. Saunders. This classic textbook is written for students of radiation oncology and nuclear medicine, with much of the material relevant to medical exposures. It is also a useful text for anyone with interest in radiation biology.

Hall, E. J. 2000. *Radiobiology for the radiologist,* 5th ed. Philadelphia: Lippincott Williams & Wilkins.

Armed Forces Radiobiology Research Institute (AFRRI). 2003. *Medical management of radiological casualties,* 2nd ed. Bethesda, MD. This is a small, concise handbook aimed at health care professionals, especially those in the military, with information on radiation health effects and medical management of patients; it can be downloaded free of charge (www.afrri.usuhs.mil/outreach/pdf/2edmmrchandbook.pdf).

Committee for the Compilation of Materials on Damage Caused by the Atomic Bombs in Hiroshima and Nagasaki. 1981. *Hiroshima and Nagasaki. The physical, medical, and social effects of the atomic bombings.* E. Ishikawa and D. L. Swain, trans. New York: Basic Books. This is a detailed and unique account of human health effects from the 1945 atomic bombings compiled by the Japanese authorities and translated to English.

For an in-depth review and discussion of the current knowledge about radiation health effects at low doses, see the following three sources:

National Academy of Sciences, 2006. Health risks from exposure to low levels of ionizing radiation. BEIR VII report. Washington, D.C.: National Academies Press (www.nap.edu/openbook.php?isbn=030909156X).

French Academy of Sciences (Académie des Sciences). 2005. Dose–effect relationships and estimation of the carcinogenic effects of low doses of ionizing radiation. This report was prepared in collaboration with the French National Academy of Medicine

(Académie Nationale de Médecine). The English version can be downloaded from www.academie-sciences.fr/publications/rapports/pdf/dose_effet_07_04_05_gb.pdf

International Commission on Radiological Protection (ICRP). 2006. *Low-dose extrapolation of radiation related cancer risk.* ICRP report no. 99. New York: Elsevier.

On the topic of pregnancy and radiation, the following two sources provide in-depth information, especially for professionals dealing with prenatal radiation exposure issues:

Wagner, L. K., R. G. Lester, and L. R. Saldana. 1997. *Exposure of the pregnant patient to diagnostic radiations: A guide to medical management.* 1997. Madison, WI: Medical Physics Publishing.

ICRP. 2000. *Pregnancy and medical radiation.* ICRP report no. 84. New York: Elsevier.

A number of brief and informative physician fact sheets on acute radiation syndrome, cutaneous radiation syndrome, and prenatal radiation exposure can be downloaded from the CDC Web site (http://emergency.cdc.gov/radiation/clinicians.asp).

The Health Physics Society's "Ask the Expert" feature offers a number of informative questions and answers on the topic of pregnancy and radiation for the general public (www.hps.org/publicinformation/ate/cat4.html).

7

ENVIRONMENTAL EFFECTS OF RADIATION

OVERVIEW

IMMEDIATELY FOLLOWING A RADIATION INCIDENT, PEOPLE take action to protect themselves and their families, save lives, and help others. We will discuss response to radiation emergencies in Part Three, "Responding to Radiation." Once the initial impact of a radiation disaster is over, however, many questions remain with varying degrees of urgency. If homes have been evacuated, when is it safe to return? When is it safe for area businesses to start operating? Have the local sources of drinking water been affected? Can farmlands be cultivated and dairy farms operated? When is it safe to walk in the park, play in the yard, fish in the lake, or hunt in the woods? In other words, when can life go back to normal?

The answer is dictated in part by the overall environmental impact of the radiation incident and how that incident has affected air, land, water, and infrastructure. In a purely technical sense, this situation is no different from any other technological disaster where, for example, oil or chemical toxins may affect the environment. However, when radioactivity is involved, perceptions are an additional factor to overcome.

In this chapter, we describe how radioactivity, in general, can adversely affect the environment and what can be done to mitigate the radioactive contamination. Then we describe the range of environmental effects to be expected from the radiation scenarios described in Chapters 4 and 5 and provide examples from past incidents. In short, some types of radiation accidents may have limited or no environmental impact or they may only affect a small local area. Other types of radiation incidents can have devastating effects on the environment. If a large amount of radioactive material is released into the air, a serious inhalation hazard will persist for some time as radioactive particles disperse in the air or the wind kicks up radioactive dust that has already settled on the ground. Building surfaces and roads can be contaminated. Much of that contamination can be washed away, but some may be difficult to clean. Some buildings may have to be demolished.

Contamination in farmlands may render crops from those areas unsuitable for consumption. Contaminated vegetation can affect the meat and milk of grazing animals. Health and environmental professionals monitor radioactive contamination in the environment and foodstuffs to assess whether an area is suitable for return to occupancy or farming as appropriate. People can live safely in areas under agricultural restrictions as long as they observe the health advisories against using certain locally grown foodstuffs.

Recovery from any and every radiation contamination incident can be achieved. The recovery may be costly and it may take time. The pace of this recovery will depend, to a large extent, on how the authorities earn and maintain the public trust and how rational and pragmatic the public remains in its approach to recovery and a return to "normalcy."

AFFECTING THE ENVIRONMENT

As described in Chapter 2, radioactivity is an inseparable part of the natural environment that is present in what people breathe, eat, and drink. A radiation incident can introduce *additional* amounts of radioactive materials into the environment. The true impact on health and way of life depends on the amount of additional radioactivity and how it is introduced into and interacts with the environment. Radioactive materials can contaminate the air, building surfaces, soil, surface water, and groundwater; they can also enter the food chain from multiple pathways. In this section, we broadly review these exposure pathways. Later in the chapter, we discuss specific examples.

Air

Radioactive materials may release into the air in the form of gas or particles. This cloud of radioactive materials is called the "plume" and it may or may not be visible. Radioactive gases may be emitted from a nuclear power plant. Some radioactive gases, such as xenon or krypton, are inert, have short half-lives, dissipate quickly, and do not present a long-term environmental hazard. Radioactive iodine is also released in the form of gas, but it can readily combine with other chemicals and linger in the environment longer. Radioactive materials can also contaminate the air in the form of small suspended particles. The sources of such radioactive particles can be

- uncontrolled release from a smoke stack of an industrial facility where a radioactive source is accidentally smelted;
- uncontrolled release of fission products from a nuclear power plant;
- detonation of a dirty bomb or another form of radiological dispersal device; or
- detonation of a nuclear device.

The radioactive particles released into the air are blown by the wind, and ultimately settle onto the ground surface. The science of dispersion and deposition of radioactive particles is complex. However, generally speaking, the extent of environmental damage is influenced by how much radioactive material is released, the physical and chemical characteristics of the contaminated particles, how far into the air the particles are lifted or ejected, the prevailing winds at various altitudes, the terrain underneath, and whether or not it rains.

Building Surfaces

Radioactive particles that land on external surfaces of buildings, sidewalks, and streets contaminate those surfaces. Radioactive materials may also enter buildings through ventilation or by people tracking them inside. Much of this contamination can usually be removed simply by wiping or washing. Some of the contamination may be more difficult to clean because the radioactive material may be lodged in cracks and crevices. A variety of surfaces exists in urban environments, such as concrete, asphalt, granite, marble, brick, steel, glass, wood, and plastic. Some radioactive materials may chemically interact with the surface material and become attached to those surfaces. Harsher cleaning techniques (such as sandblasting or scabbling) may be required to clean some surfaces. In technical jargon, surface contamination that can be cleaned by simply wiping or washing is said to be "removable," "mobile," or "loose" contamination. Whatever amount of radioactivity remains afterward is said to be "fixed" contamination.

Radioactive contamination that is fixed to a surface can be problematic because it is harder to clean. However, fixed contamination offers a distinct safety advantage because it cannot be moved, scattered (resuspended) in the air, or get onto skin or clothing and possibly be inhaled or ingested. The health implications of surface contamination depend, therefore, on what fraction of the radioactive material is fixed and what fraction is mobile or removable. The fixed contamination can still be a hazard if someone stays close to the contaminated surface for a length of time. The total amount and the type of radioactive material, as ascertained by trained professionals, are important determinants of how dangerous or benign the contamination may be at that location and for how long.

In the aftermath of a radioactive contamination event, any loose contamination on roadways has a potential to become airborne. Rain is helpful in this regard. Spraying water to keep the dust down is an option. Authorities will need to prioritize contaminated areas that need to be cleaned. Areas with higher levels of contamination or areas more accessible to the public, especially to children, have higher priority for cleanup. Critical infrastructures (hospitals, utility buildings, financial institutions, etc.) also have priority so that operations can resume in a safe environment as quickly as possible.

There are other considerations as well. For example, if the radionuclide of concern has a short half-life (Chapter 3), it may be more prudent to cordon off the area and let the radioactivity decay to acceptable levels by itself. If the radionuclide is an alpha emitter, trying to decontaminate the material may remobilize the contaminant and create a serious inhalation hazard for the cleanup crew. In that case, it may be more prudent, as an interim measure, simply to apply a layer of fresh paint over the surface to "fix" the contamination in place and also shield the alpha particles. If such measures are taken, the painted areas need to be monitored periodically to make sure the paint is not flaking off the surface. This monitoring should continue until those surfaces can be permanently cleaned or the radioactivity decays to acceptable levels.

From a long-term perspective, widespread *fixed* contamination on street surfaces and building walls could present a public perception issue even if actual radiation levels are low.

Soil

Radioactive particles that settle on soil will contaminate it. The contamination will be primarily limited to surface layers of soil, but with time and weathering, some of the contamination will migrate downward to subsurface layers. At the surface, the particles may become airborne by wind and pose an inhalation hazard. This hazard is especially important with radionuclides that emit alpha particles. Contamination in the top layers of soil can also be taken up by roots of vegetation in that area. Over time, the radioactive contamination in soil can be mobilized by rain, groundwater, and erosion.

Excavating land areas containing contaminated soil can quickly amount to large volumes of soil. Finding a permanent storage place for the radioactive waste thus generated may not be easy. If large areas of soil are affected, excavation and removal of contaminated soil will become cost prohibitive and may not even be practical. Alternatively, land use can be restricted and methods can be employed to limit erosion, such as planting trees or maintaining vegetative cover. Deep plowing or tilling the land, which mixes the surface soil with soil layers underneath, can be a practical approach to lower the concentration of radioactive materials over large areas of land.

The International Atomic Energy Agency (IAEA) considered 20 years of research and experience following the Chernobyl accident and offered a simple and practical alternative to reduce exposure to radioactive contamination in soil.* IAEA suggested removal of the upper 5–10 cm layer of soil (the most contaminated layer) from residential yards and around public buildings, schools, and kindergartens, and placing the waste into holes specially dug on the same properties. The clean soil excavated from the holes can be used to cover the decontaminated areas. This approach does not require transportation and disposal of contaminated soil in special burial sites. Although this recommendation seems practical, it is unlikely to receive public acceptance in most communities.

Water

Bodies of water such as lakes and rivers can become contaminated through direct deposition of radioactive material from the air, erosion from neighboring contaminated soil, or direct discharge of contaminated fluids into a lake or river. In all these cases, dilution works to our advantage.

Consider a plume of radioactive material that passes over a large lake and deposits radioactive materials on the lake surface. As this surface contamination mixes with the large volume of water, the concentration decreases significantly. Such bodies of contaminated water are seldom an immediate hazard to people. After the plume has passed, the water itself presents no danger to navigation.† However, restrictions will likely be placed over swimming and fishing in these waters for some time. The sediment at the bottom will have measurable levels of radioactivity as the material settles to the bottom. Removal of contaminated sediment is seldom practical and presents the same logistical challenges as removal of contaminated soil from large areas of land.

Groundwater can become contaminated depending on the chemical nature of contamination, its mobility through soil, and the location of aquifers. The processes employed in water treatment facilities can remove a large amount of contaminated particles from water. This process results in high concentrations of radioactive sludge at the treatment facility. Any dissolved radioactivity in the water is not filtered. Therefore, when there is potential for a radiation incident to contaminate sources of drinking water, the water needs to be monitored at the source or as it enters the water treatment facility.

---

* International Atomic Energy Agency. 2006. Environmental consequences of the Chernobyl accident and their remediation: Twenty years of experience. Vienna: IAEA.

† Winds blowing from a contaminated shoreline may carry radioactive dust and still pose a danger to those navigating the waterway.

Alternate sources of drinking water should be considered if concentrations of radioactive materials exceed allowable limits.

Vegetation

As with any other surface, radioactive materials will settle on surfaces of vegetation and contaminate them. If the settling is by "dry deposition" (i.e., no rain), grassy areas and trees with leaves will retain much more of the radioactive deposits than paved streets in urban areas. Rain can wash away much of this contamination into the ground. In addition, as noted earlier, radioactive materials deposit directly on the ground and contaminate the soil. As the roots absorb the radioactive material from the contaminated soil, the trees and vegetation become contaminated. A heavily contaminated tree may have to be removed. On the other hand, if large areas of soil are affected by contamination, leaving trees and vegetation in place helps control erosion and spread of radioactivity in the soil, even though the trees and vegetation are contaminated.

> *Did you know?* The large and sticky leaves of tobacco plants attract and collect naturally occurring radioactive materials in the air originating from radon (Chapter 2). These include polonium-210, which attaches to the tobacco leaves and later becomes volatilized when the cigarette is lit; it is inhaled along with the smoke. This inhaled polonium is believed to be an important factor in initiating lung cancer among smokers.

AFFECTING THE FOOD CHAIN

Radioactive contamination can deposit on the outside surfaces of foods and crops, or it can become incorporated through metabolic uptake in plants, fruits, vegetables, and animals and enter the food chain. The former (surface deposition) is a short-term concern in the first few days and weeks after the incident. The latter (metabolic uptake by vegetation and animals) is a more long-term health concern.

If radioactive materials are released into the environment, authorities will issue advisories in the affected area to warn people against using vegetables from their home gardens or consuming any locally grown fruits and vegetables until the health authorities have made measurements to assess the levels of radioactivity in those products. If already harvested fruits and vegetables are stored uncovered when the radioactive plume passes over them, the surfaces of those products will be contaminated. If they must be used, they should be washed thoroughly and peeled if possible. Leafy vegetables from home gardens are a special concern because they have a larger surface area on which contaminated particles can land. It is more prudent to avoid consuming those products if alternate sources of food are available.

Crop surfaces can become contaminated in other ways. For example, raindrops hitting the soil can deposit small contaminated particles on plants by splashing. The mechanical action of harvesting can also transfer contaminated particles from soil to the crop. However, a more serious type of crop contamination, which cannot be washed away, is through root uptake from the soil.

Radioactive contamination in the soil can pass to the roots and from roots to the edible portions of vegetation. As grazing animals consume this contaminated vegetation, their meat and milk become contaminated. Human diet will be affected through consumption of contaminated vegetation or animal products. Many factors can influence this process along the way, including chemical properties of the contaminant, its availability in the root zone for plant uptake, and subsequent metabolism of grazing animals. The contaminant pathway from soil to grass to a cow (or goat) and to the milk occurs very rapidly. Children drinking locally produced contaminated milk are a special concern because they are more sensitive to the health effects of contaminated foods.

Although the largest impact on the food chain is through contaminated soil affecting farming and livestock, contaminated bodies of water can also affect the food chain through what is technically referred to as the "aquatic pathway." Fish swimming in contaminated waters have a tendency to concentrate radioactive materials in their bodies. In fact, researchers often analyze and use samples of fish to determine whether a body of water has low levels of contamination. It is best to refrain from eating fish from suspect sources until local health advisories are lifted.

Putting It in Perspective

We should put the discussion of a contaminated food chain in perspective. When the source of food contamination is environmental (e.g., meat from a contaminated cow or deer, vegetables grown on contaminated land, or milk from a contaminated goat or cow), *limited servings* of these foods will certainly not be life threatening and may not even increase the risks of long-term health problems to any significant extent. It is the *continued* diet of such foods that could place health at serious risk. Children, of course, are especially vulnerable. A continued diet of contaminated milk can have serious health consequences for children. But single or limited servings of such indirectly contaminated foods will not place their health in jeopardy.

The reason for providing this perspective is that if you or your children unknowingly consume single or limited servings of such foods before being alerted to the possibility of their contamination, it is not necessarily a cause for alarm. Furthermore, in unusual circumstances when an alternative food supply is not available, starvation or dehydration has far more immediate health consequences than consumption of food with environmental levels of radioactive contamination.

RANGE OF EFFECTS

We have explored a variety of ways that the habitat can become contaminated with radioactive materials. The environmental impact resulting from radiation emergencies has a wide range. To illustrate this range, let us examine the likely environmental impacts of various accidental and intentional scenarios discussed in Chapters 4 and 5.

Nuclear Power Plants

An accidental release of radioactivity from a nuclear power plant can have adverse environmental effects over a large area and for a long period of time, especially affecting the agricultural and farming industries. As a result of the 1986 Chernobyl accident, an area of approximately 150,000 km^2 (58,000 $mile^2$) became contaminated in the former Soviet Union.* This area included 20% of the entire Belarus territory. Another 45,000 km^2 (17,000 $mile^2$) was contaminated in northern and eastern Europe. The Chernobyl accident resulted in contamination of crops, cattle, and dairy in large geographic areas (Figures 7.1 and 7.2). Years after the accident, crops in more highly contaminated areas still contain levels of radioactivity with higher than recommended limits for consumption.

It is noteworthy that millions of people can live safely in areas that are still considered contaminated as long as they observe health advisories against using certain locally grown foodstuffs. The highest concentrations of cesium-137 are found in food products from forest areas, especially in mushrooms, berries, wild game, and reindeer. High concentrations of cesium-137 in fish occur in lakes with slow or no turnover of water, particularly if the lake is also shallow.

Socioeconomic factors have placed some of the population in rural areas at a disadvantage. Some of these people have found it difficult to afford the "clean" products brought from unaffected regions. Therefore, they ignore government advisories and consume the restricted locally grown foodstuffs. Locally produced products such as milk are not routinely tested

* Any land area with a cesium-137 deposition greater than 37 kBq/m^2 (1 Ci/km^2) was designated as contaminated (see the United Nations 2000 report in the "Resources" section).

Figure 7.1

An environmental team evaluating radiation levels and taking grass samples near Polesskoje, Ukraine, in 1990. This effort was part of the IAEA International Chernobyl Assessment Project in which approximately 200 scientists from 25 countries participated. (Photo: IAEA.)

for radioactivity and may contain levels above recommended values.* It has been shown that boiling meats and vegetables and discarding the broth before cooking can potentially reduce the amount of radioactivity in the food, but the results are variable and the prudent action is to avoid consumption of restricted food products.

Elevated levels of radioactivity have been detected in bodies of people consuming restricted foods and estimates of their radiation doses are available.†

* The radionuclides of concern are cesium-137 in most of the food products and strontium-90 in milk. Because iodine-131 has a short half-life, it is no longer an environmental or dietary concern in these areas, but in the first few weeks after the Chernobyl accident, children's consumption of milk contaminated with radioactive iodine contributed to an increase in the rate of thyroid cancer (see Chapters 4 and 6).

† For example, see Travnikova, I. G., A. N. Bazjukin, G. Ja. Bruk, V. N. Shutov, M. I. Balonov, L. Skuterud, H. Mehli, and P. Strand. 2004. Lake fish as the main contributor of internal dose to lakeshore residents in the Chernobyl contaminated area. *Journal of Environmental Radioactivity* 77:63–75.

Figure 7.2
Buying milk samples from local farmers to analyze content for radioactivity. This effort was part of the IAEA International Chernobyl Assessment Project in which approximately 200 scientists from 25 countries participated. (Photo: Elisabeth Zeiler/IAEA.)

However, it remains to be seen whether any long-term health consequences in this population can be directly attributed to consumption of radioactively contaminated foodstuffs.

Did you know? Norway used the drug Prussian blue (see Chapter 12) as a feed additive to treat thousands of cesium-contaminated sheep, reindeer, and cattle and achieved significant reduction in the levels of radioactivity. This successful program was then implemented in Russia, Belarus, and Ukraine to treat livestock and was successful in reducing the levels of cesium-137 contamination in meat, milk, and crops fertilized with the animals' manure.*

Did you know? Cesium-137 contamination from Chernobyl settled on certain upland areas of the United Kingdom where sheep farming is the primary land use. As a result, severe restrictions were placed on the movement, sale, and slaughter of sheep from the affected regions in England, Wales, Scotland, and Ireland; this initially affected nearly 9,000 farms. These restrictions lasted for several years and have now been lifted from the majority of the farms as the radioactivity levels have fallen. But many farms are still operating under the restrictions. Sheep from those regions can be moved to clean areas for several weeks of grazing before being tested and sent to slaughterhouses.†

Did you know? A group of Irish and Ukrainian researchers found that curing radioactively contaminated meat in a sodium-chloride salt solution (brine) for an hour to a day could eliminate 50–80% of the cesium-137 contamination.‡ This finding is only an academic curiosity—not a recommended approach to try at home—and it was never used commercially.

In sharp contrast to the situation after Chernobyl, thousands of environmental samples of air, water, milk, vegetation, soil, and foodstuffs that were collected in 1979 from the Pennsylvania countryside surrounding the Three Mile Island reactor showed very low levels of radioactivity, and the environmental impact from that accident was virtually zero. This fortunate outcome was because the vast majority of radioactivity at Three Mile Island was contained within the reactor and did not release to the outside environment.§

Did you know? The Chernobyl accident released 1,760 PBq (50 million Ci) of iodine-131 and 85 PBq (2 million Ci) of cesium-137 to the environment.¶

* According to a 2005 report by the International Atomic Energy Agency, an annual investment of US$50,000 in the Prussian blue treatment program saved an estimated US$30 million worth of milk and meat production a year in Belarus (www.iaea.org/NewsCenter/Features/Chernobyl-15/agriculture.shtml).

† The radiation monitoring program called "Mark and Release" applied a criterion of 1,000 Bq of cesium-137 per kilogram of body weight (1,000 Bq/kg) to certify that the meat was clean. Any sheep exceeding the limit was marked with a dye for later evaluation.

‡ Long, S. C., S. Sequeira, D. Pollard, et al. 1996. The removal of radionuclides from contaminated milk and meat. In The radiological consequences of the Chernobyl accident, 519–523. Brussels, Luxembourg: European Commission, EUR 16544 EN.

§ Report of the President's Commission on the Accident at Three Mile Island. 1979. Washington, D.C. Also see the Nuclear Regulatory Commission fact sheet on the Three Mile Island Accident (available from www.nrc.gov/reading-rm/doc-collections/fact-sheets/3mile-isle.html).

¶ A pitabecquerel (PBq) equals a million billion becquerels (10^{15}).

In comparison, the accident at Three Mile Island released less than 0.0007 PBq (20 Ci) of iodine-131 and no cesium-137.*

Dirty Bomb

The environmental impact of a dirty bomb explosion depends on the specific circumstances. If the bomb is detonated in an urban area and the radioactive material is in a form that could be volatilized and dispersed, several blocks of the city could become contaminated with potentially significant amounts of radioactivity. A much larger area, miles away from the scene, may also have *detectable* amounts of radioactivity. Important factors at play are the amount of radioactive material used in the bomb, the design of the explosives, the surrounding terrain, and the weather conditions (i.e., winds at different altitudes and presence of precipitation). If detonated in an urban area, the challenging and costly aspect of recovery will be the cleanup of contaminated and damaged buildings, streets, and sidewalks. It may be difficult or cost prohibitive to clean some surfaces to pre-incident baseline levels. In other words, some residual amounts of fixed radioactivity may remain. (See "Decisions for the Long Term" later in this chapter.)

Any agricultural impacts from a dirty bomb incident are likely to be short term. There are several reasons for this assertion. The radioactive material used to construct a dirty bomb may have high levels of radioactivity, but it is still limited in size (i.e., the amount of radioactivity will be orders of magnitude less than that in the Chernobyl accident). If the bomb is designed for maximum dispersion, the radioactive dust will contaminate a larger area; however, the extensive dispersion will dilute the radioactivity, so contamination will be at lower concentrations. Besides, the bomb is likely to detonate in a large metropolitan area far from any farms or agricultural fields. Nevertheless, health authorities need to monitor the environment vigorously and take prudent precautionary measures until the environmental monitoring results become available.

Nuclear Detonation

Chapter 5 described the effects of a nuclear detonation and the difference made by the explosive yield of the weapon as well as the height of the

* Almost all of the radioactive iodine (20 million Ci) from the reactor fuel at Three Mile Island was contained in the primary system and containment and auxiliary buildings and was not released to the outside. Similarly, a considerable amount of cesium-137 was retained inside the reactor containment, and no detectable amounts escaped to the environment.

explosion (air or ground burst). In addition to the immediate devastation and impact on human life, those factors also influence the extent of environmental impact from the blast. As we discussed in Chapter 5, if a nuclear weapon is detonated near the ground surface (the likely terrorist scenario), the fireball touches the ground; large amounts of soil and building materials are vaporized and sucked into the mushroom cloud and later settle down as nuclear fallout. This fallout will be the source of environmental contamination at distances far from the blast site.

The largest fallout particles, as well as the most highly radioactive, generally fall within the first few hours and weather conditions (wind and precipitation) determine deposition patterns. Smaller particles are dispersed into the troposphere, stay aloft longer, and travel longer distances before they, too, settle to the ground in a matter of days. With high-yield weapons (greater than 500 kilotons), debris will penetrate the stratosphere and over a period of many months will deposit the radioactivity on a global scale. The longer it takes for the material to settle, the lower the radioactivity levels will be and the more dispersed the material will become. The benign levels of radioactive cesium that can be measured in backyard soil today are remnants of atmospheric nuclear weapons testing from 60 years ago.

> *Did you know?* Levels of radioactivity in debris from a nuclear explosion drop 20-fold from the first hour to 24 hours after the explosion and 100-fold by 48 hours after the event.

A nuclear weapon constructed by a terrorist organization is likely to have a yield much less than 100 kilotons and therefore the debris will not reach the stratosphere. But the environmental effects could still be measured thousands of miles away. The atmospheric weapons testing done at the Nevada test site were in the kiloton range—the range of yield expected from an improvised nuclear device (IND) in a terrorism incident. The very first atomic weapon detonated in the New Mexico desert was a 19-kiloton device detonated at a height of 30 m (98 ft) (i.e., ground burst) with the fireball touching the ground and producing a significant amount of fallout.

The pattern of fallout deposition can be very spotty. For example, after an atmospheric test in Nevada in April 1953, the highest levels of fallout, except for the vicinity of the bomb site, were measured in Troy, New York—2,500 miles (4,000 km) away—because of rain. That incident caused measurable radiation doses, but it did not present an immediate health risk to the public.

Therefore, outside the area that is immediately devastated by a nuclear detonation, nuclear fallout can cause a range of long-term environmental issues similar to what was described for the Chernobyl accident. Areas of contaminated soil can be expected. There will be areas where sources of drinking water may become contaminated as a result of direct deposition of fallout or water runoff. In some areas, agricultural products will be affected and dietary restriction advisories will remain in effect for decades.

Other Scenarios

The scenarios we discuss in this section occur with some frequency throughout the world. They are more probable than the scenarios discussed earlier.

Lost or Orphan Sources As long as the integrity of a radioactive source is not breached (i.e., the radioactive material does not leak or spread), there will be no contamination and the environmental impact will be zero. (See the 1999 Samut Parakan accident in Thailand discussed in Chapter 4.) However, if the integrity of the radioactive source is compromised and the source is spread, environmental consequences will occur. The 1987 Goiânia accident in Brazil contaminated 46 homes and cleanup activities generated 3,500 m^3 (4,600 yard^3) of radioactive waste (Figures 7.3 and 7.4). To put these types of accidents in perspective, their environmental impact is limited to a small geographic area; although it creates extreme hardship for the people involved, its direct impact affects a relatively small population.

Figure 7.3
Removing contaminated debris as part of the decontamination efforts in Goiânia, Brazil. (Photo: IAEA.)

Figure 7.4
Approximately 3,500 m^3 (4,600 yard^3) of contaminated waste is buried in these two grassy hills on the outskirts of Goiânia. This picture was taken in October 2007, 20 years after the accident. (Photo: Kirstie Hansen/IAEA.)

Industrial Accidents Industrial accidents could result in discharge of radioactivity through a smoke stack. Radiation dose to individuals is likely to be low, but the surrounding environment will be affected and cleanup costs could be significant. If the accidental discharge of radioactivity is to the sewer, the sewage treatment plant is likely to become contaminated. Overall, the environmental impact of these types of accidents tends to be limited to a small geographic area and cleanup will be practical and manageable, albeit costly.

Transportation Accidents If a container of radioactive material being transported is not breached, there will be no contamination and no environmental consequence as far as radioactivity is concerned. It is extremely unlikely that containers of high-level radioactive material would be breached as a result of transportation accidents, even if fire were involved. However, medical radioisotopes with much lower amounts of radioactivity are routinely shipped and these containers can be compromised in a transportation accident. If radioactive material is released from the containers, some of the radioactivity may be dispersed to some distance; however, the concentrations will most likely be low and environmental impact will be minimal or nonexistent away from the scene of the accident. Furthermore, the half-lives of medical isotopes are short and any contamination will decay soon.

Medical Administration Environmental impact from medical administration of radioactive materials is very limited. Patients who are administered radionuclides and released from hospitals leave residual radioactive contamination at public places they visit, and the sanitary products they use occasionally set off radiation alarm monitors at landfills. Because the radiation levels are low and the half-lives of medical radioisotopes are short (measured in hours or days), the environmental impact is zero to minimal.

DECISIONS FOR THE LONG TERM

During normal times, strict regulatory limits are placed on the levels of radioactivity entering the environment as a result of commercial, industrial, or research activities. Adhering to these limits is part of the regulatory requirements for any entity holding a federal or state license to work with radioactive materials. If the environment becomes contaminated as part of these operations, guidelines for cleaning it can be followed. However, in the case of a serious radiation emergency resulting from a dirty bomb or nuclear detonation, there are no preestablished guidelines to apply to all areas needing to be cleaned.

What are the appropriate cleanup criteria and methodologies and how are the most suitable ones selected? The answer depends on a variety of technical and societal factors. These include the size, type, and location of contamination, human health risks, projected land use, economic effects (e.g., on residents, tourism, commerce), preservation or destruction of places of national or historic significance, technical feasibility of cleanup, potential adverse effects of remedial actions, ecological risk, cost effectiveness, waste disposal options, cultural sensitivities, and overall public acceptability. It should be clear that the "right" answer is not simply a technical answer and that no single person or entity can make this judgment.

Cleanup levels that may be suitable for an industrial warehouse building may not work for a bank, and cleanup levels suitable for a bank may not be acceptable for a residential community or an elementary school. Cleanup methodologies that may be suitable for remediation of the CNN Center in downtown Atlanta may not be appropriate for the Ebenezer Baptist Church, 1 mile away, or the residential homes nearby. Although it is practical to remove all contaminated soil from the National Mall area in Washington, D.C., it is not practical to do so from a hundred square miles of Virginia farmland.

Unlike the early decisions in an emergency, which are made in a time-sensitive manner by responsible authorities, decisions about the long-term environmental cleanup of contaminated areas need to be made by a deliberative process involving all stakeholders. The stakeholders include, at a minimum, people whose health, quality of life, and economic well being will be affected by the decision and those who have regulatory responsibility. A sound cleanup approach that would be protective of human health and environment can then be decided with all technical and scientific information taken into account, together with societal objectives, priorities, and cost factors.

Resources

The following is a classic textbook offering technical information on the subject of environmental radioactivity and transport of radioactive materials in the environment. Chapter 12 of this reference provides an overview of several radiation contamination accidents, such as the 1957 Windscale (Sellafield) nuclear reactor accident in northwest England, the 1978 reentry of Soviet satellite *Cosmos 954* into Canadian territory with a nuclear reactor onboard, and the 1966 crash of two U.S. Air Force aircraft over the southeastern coast of Spain, which resulted in four hydrogen bombs falling from one of the planes (no nuclear detonation occurred) and spreading of plutonium contamination over a square-mile area:

Eisenbud, M., and T. Gesell. 1997. *Environmental radioactivity from natural, industrial, and military sources.* San Diego: Academic Press.

The following reference is U.S. government planning guidance with detailed background information on protection action guides. It also elaborates on the importance of stakeholder involvement in the decision-making process for long-term recovery:

Department of Homeland Security. 2008. *Planning guidance for protection and recovery following radiological dispersal device (RDD) and improvised nuclear device (IND) incidents.* 73 FR 45029.

The following three references contain more detailed information about the environmental consequences of the Chernobyl accident:

International Atomic Energy Agency. 2006. *Environmental consequences of the Chernobyl accident and their remediation: Twenty years of experience.* Vienna: IAEA.

United Nations Scientific Committee on the Effects of Atomic Radiation. 2000. Annex J: Exposures and effects of the Chernobyl accident. In *Sources and effects of ionizing radiation: United Nations Scientific Committee on the Effects of Atomic Radiation UNSCEAR 2000 report to the General Assembly, with scientific annexes, volume II: Effects.* New York: United Nations (available from www.unscear.org/docs/reports/annexj.pdf).

European Commission. 1996. *The radiological consequences of the Chernobyl accident.* EUR 16544 EN, Brussels, Luxembourg. This publication contains the proceedings of an international conference in Minsk, Belarus, and offers many technical papers on environmental, health, and medical aspects of the Chernobyl accident.

For readers who are interested in methods of processing foodstuffs to eliminate radioactivity, the following reference is somewhat dated, but provides a useful overview of the literature:

Noordijk, H., and J. M. Quinault. 1992. The influence of food processing and culinary preparation on the radionuclide content of foodstuffs: A review of available data. In *Modeling of resuspension, seasonality and losses during food processing.* TECDOC-647, 35-59, International Atomic Energy Agency: Vienna (available from www-pub.iaea.org/MTCD/publications/PDF/te_647_web.pdf).

8

PSYCHOSOCIAL EFFECTS OF RADIATION

OVERVIEW

When radiation disasters are discussed, the tendency is to focus on the potential physical destruction and bodily harm they may cause as well as any long-term health or environmental impacts of such disasters. However, the significance of psychological and social effects of such traumatic experiences should not be underestimated. The status of mental and behavioral health can affect, even impair, the capacity to respond to an unfolding disaster and influence the life and death decisions that may need to be made. Experiencing emotional trauma can also cause serious long-term psychological harm for disaster victims as well as professional responders. At a minimum, awareness of disaster mental health issues is essential for all, especially those participating in disaster response.

Acts of terrorism further exacerbate the psychosocial impact on affected populations. The threat of radiation and radioactivity will also be an additional stressor in any disaster scenario because people regard radiation as a horrifying hazard. Let us consider a dirty bomb scenario (Chapter 5). Most experts agree that such a radiological terrorism incident is better described as a weapon of mass *disruption* (not destruction). In such a scenario, the extent of physical damage is limited, but radiation is used as a tool for psychological warfare to create terror. It goes without saying that the true impact of a dirty bomb incident is determined ultimately by how people react to it.

Some individuals may consider mental and behavioral health issues as inevitable consequences of disasters and therefore not regard them as a high priority, especially in the immediate aftermath of a disaster. The fact is that through appropriate training of responders and early intervention practices, mental health issues and long-term psychological and social impacts can be mitigated to a significant extent. In the face of a catastrophic *nuclear* incident, mental and behavioral health influences how quickly recovery takes place. Therefore, resiliency to any catastrophic incident is measured not only by the strength of the infrastructure, but also by mental health and how individuals, communities, and the nation respond to such incidents.

This chapter describes the psychosocial effects of radiation disasters, provides a number of real-life examples, describes signs and symptoms to look for in families or co-workers, discusses useful approaches to mitigating these effects, and suggests resources for additional information and training.

DEFINING THE IMPACT

The term "psychosocial" refers to the psychological and mental health of disaster victims and emergency responders and their interplay with the surrounding social conditions and environment during and after a traumatic experience. The signs and symptoms of traumatic stress can take different forms and vary according to the nature and severity of the experience as well as the individual risk factors. For emergency responders, it helps to be able to recognize these symptoms in themselves, their colleagues, and the victims and their families whom they are trying to help. Common manifestations of stress situations can be grouped into the following categories:

- Cognitive: Symptoms include disorientation, confusion, memory loss, and impaired ability to concentrate, think clearly, or solve problems.
- Physical: Symptoms include fatigue, insomnia, changes in appetite, nausea, vomiting, headaches, bodily aches and pains, or rash. Some of these physical stress reactions mimic symptoms of major disorders or disease. For example, nausea and vomiting are also signs of excessive radiation exposure (Chapter 6).
- Emotional: Symptoms include depression, anxiety, numbness, guilt, fear, pessimism, anger, resentment, hopelessness, or helplessness.
- Behavioral: These reactions may take the form of social withdrawal or isolation. They may also take the form of increased drinking, smoking, and substance abuse in general; irritability; mistrust; and aggression.
- Spiritual: Individuals may question the meaning of life or question their faith.

Every individual who experiences a traumatic event is likely to develop one or more of these symptoms. For many, the signs and symptoms improve with time. But for some individuals, the emotional impact of the experience may develop into posttraumatic stress disorder (PTSD) and persist for months, years, or a lifetime after the incident.

If people are forced to relocate from the disaster area, additional societal and economic pressures are experienced. For example, the host community may show some resentment, and the displaced families have to find new schools for their children and struggle to find new jobs to support themselves.

Radiation as an Additional Stressor

A radiation incident, especially if it is an act of terrorism, exacerbates the psychosocial effects typically seen after natural disasters. Some of the reasons for this heightened feeling of despair include:

- Radiation is scary. Most people lack a rudimentary knowledge of radiation properties and its effects and have no experience dealing with that threat on any scale, so radioactivity is an unfamiliar threat for them.
- There is no sensory perception for the hazard. The presence of radiation cannot be seen, heard, smelled, tasted, touched, or sensed. Therefore, there is uncertainty about exposure.
- If the incident is an act of terrorism, it is likely to occur without any warning or time to make preparations. The malicious intent of a terrorism incident, its unpredictability, and concerns about further acts of terrorism add to the already heightened sense of vulnerability.
- The perception is that once someone is exposed to radiation, the exposed individual is doomed.
- The long-term risks of radiation exposure (i.e., cancer) and associated uncertainty are another source of anxiety that results in a continued sense of vulnerability. After the incident is over, the worries about possible future health effects persist for a long time. These concerns may carry to next generations as well. Japanese survivors of atomic bombs worried about the health of their children and grandchildren.
- The victims may continue to associate the mildest everyday injuries or sickness with radiation exposure, a phenomenon called "psychosomatic bind."*
- These stress reactions can occur even in the absence of actual exposure to radiation. Research has shown that people who are unexposed to radiation, but have concerns that they may have been exposed, suffer from the same psychological and behavioral effects as people who have been exposed.
- People who have been contaminated with radioactivity or exposed to radiation are likely to suffer from stigma as well. Initially, they may be shunned by others who fear touching them or being in close

* This phenomenon is vividly documented by Robert J. Lifton. 1967. *Death in life: Survivors of Hiroshima*. New York: Random House.

proximity to them. Later, they may find themselves at a disadvantage or subject of discrimination in areas such as employment, health insurance, or personal relationships.

It is striking how many of these additional stressors can be reduced or entirely eliminated by a well-informed, educated public.

Examples

The personal accounts of Japanese atomic bomb survivors provide a window into the minds of these survivors as they struggled through the first few days and later their entire lives with the memories of their experience. The feeling of guilt was clearly documented; the survivors felt guilty about and ashamed of being alive while their family members had died. They did not even like referring to themselves as "survivors" because focusing on being alive might have slighted their sacred dead.*

Did you know? Japanese atomic bomb survivors came to be known by the neutral name of *hibakusha,* which means "explosion-affected persons."

Many survivors described the numbness they felt facing the tragedy. Referring to daily cremation of the dead in a vacant lot adjacent to a military hospital in Hiroshima, a nurse working there said, "Such a thing could no longer arouse emotion. Our feelings were numb." After describing pulling dead bodies from the rubble and the river (with many corpses in gruesome physical condition) and how maggot-infested wounds of the injured were treated, a relief worker said, "[People] lose all feeling. Their minds and bodies grow numb, preventing them from thinking about what they are experiencing. All capacity for tears and grief had vanished from my being."†

Many survivors described what they saw as "hell" and their own state as "fearful bewilderment." After the initial shock was over, survivors saw that those who were well for a while later fell ill and died. The survivors felt they had a "fragile existence" as they faced an uncertain future, fearing that the same fate awaited them. The response of the Japanese government, or lack thereof, made matters worse for them. Survivors continued to experience

---

* Hersey, J. 1989. *Hiroshima*. New York: Vintage Books. The information cited here is not available in the 1946 edition of the book, but rather was included in the last chapter, which was added in the later edition.

† Sekimori, G. 1986. *Hibakusha*. Tokyo: Kosei Publishing. The quotes are translated accounts by Sakae Hosaka (nurse) and Nakaichi Nakamura (relief worker).

discrimination as they searched for employment or sought government assistance. They experienced discrimination in their personal lives as well as: People refused to marry atomic bomb survivors because they feared that women might bear unhealthy children.*

The 1987 accident in Goiânia, Brazil, provides another classic example of psychosocial effects after a radiological accident (see Chapter 4 for a detailed description of this accident). Although the contamination was limited to several families and a total of 249 people, approximately 112,000 residents of this large city asked to be monitored for radioactive contamination. This overwhelming response was partly due to citizens' health concerns because of possible contamination and partly due to the need to receive documentation to prove that they were indeed clean as the residents found themselves subject of discrimination. Hotels in other parts of Brazil refused to accommodate Goiânia residents. Some airline pilots refused to fly over Goiânia or allow Goiânia residents on their planes. Cars with license plates from the area were stoned in other parts of Brazil (Figure 8.1).†

The Three Mile Island nuclear power plant crisis in the United States resulted in no physical damage and no immediate health risks (Chapter 4). But the residents living around the plant continued to experience stress and anxiety due primarily to fears and uncertainty about possible health effects. Their anxiety was exacerbated because the accident was technologically complex, they had no control over the source of danger, and they were dependent on experts' evaluation and advice. As the crisis was unfolding, the experts disagreed, people lost trust in those responsible for the plant's operation, and anxiety increased. As people were evacuating the area, a 12-year-old boy scratched a note on his bed's headboard for people who, in some future day, might visit his house in what he and many other evacuees feared might become a radiation wasteland.‡

---

* Ishikawa, E., and D. L. Swain, trans. 1981. *Hiroshima and Nagasaki: The physical, medical, and social effects of the atomic bombings.* New York: Basic Books. This is an English translation of the book originally published in Japan in 1979. The book is well documented and informative, although not free from political statements and editorials.

† Kasperson, R. E., and J. X. Kasperson. 1996. The social amplification and attenuation of risk. *Annals of the AAPSS* 545:95–105.

‡ Houts, P. S., P. D. Cleary, and T-W. Hu. 1988. *The Three Mile Island crisis: Psychological, social, and economic impacts on the surrounding population.* University Park: Pennsylvania State University Press.

Figure 8.1
Graffiti on a wall says, "Blue flowers grow in my garden," 20 years after the accident, in the neighborhood where contaminated homes were demolished in Goiânia, Brazil. (Photo: Kirstie Hansen/IAEA.)

Did you know? During the Three Mile Island crisis, many residents of the surrounding communities saw a popular movie, *The China Syndrome*. This 1979 film was about serious safety problems at a nuclear power plant and the efforts of plant operators to cover it up. A news reporter (Jane Fonda) worked with a whistleblower engineer at the plant (Jack Lemon) to expose the cover-up. The entirely fictional film had opened just 12 days before the accident and likely influenced residents' perceptions.

People at Risk

Everyone who is directly or indirectly affected by or experiences a disaster is vulnerable to emotional distress. Among the victims, children are especially vulnerable because they are emotionally less resilient and are more susceptible to long-term psychological injury. Pregnant women, breast-feeding women, adults caring for the sick and elderly, and parents of young children are also at a higher risk because of additional concerns and responsibilities they have for their loved ones.* People with a history of mental illness and people with prior traumatic experience may also be more vulnerable.

Workers responding to a radiation disaster may worry about contaminating themselves or taking contamination home. Medical staff may experience guilt or anxiety over difficult field triage decisions. Search-and-rescue workers who encounter dead bodies are at risk of emotional distress. These concerns can be alleviated to a great degree by prior training and education.

Another important aspect of working with disaster victims relates to empathy and over-identification with the victims. All workers who come in contact with and witness other people's suffering place themselves at risk of being traumatized (Figure 8.2). The more empathetic and committed the responders are, the greater is their susceptibility to this type of secondary stress, which is sometimes called "compassion fatigue." This phenomenon is especially true for people in the "helping professions," including social workers, counselors, and mental health professionals who make personal contact and hear personal stories of disaster victims. Nurses, physicians, pastoral counselors, and law enforcement personnel are also at risk (Figure 8.3).

The clinical term for compassion fatigue is secondary traumatic stress disorder (STSD). The symptoms of STSD are similar to those for posttraumatic stress disorder and include depression, anxiety, anger, exhaustion, numbness, or judgment errors. Inflicted workers are less productive and become dysfunctional. Substance abuse is possible and workers' personal lives and relationships may be affected as well. Another related phenomenon that results from overexposure and identification with disaster victims is called "vicarious trauma": The helper experiences the trauma vicariously through the victims. The "Resources" section provides a list of suggested resources where professional responders can find additional information and guidance on this topic.

> *Did you know?* Police officers working at a gruesome scene involving fatalities may use black humor as a tool for coping with the situation. This joking is not an attempt to make light of the situation, but rather a method to keep emotional distance so that they can continue their arduous task and maintain their effectiveness.

---

* A study of a population of children in Kiev who had been evacuated from Chernobyl when they were infants or in utero was conducted. It found no differences between these children and their peers in terms of physical, social, and scholastic competence 11 years after the accident. The children perceived their well-being as similar to that of their peers. However, the evacuee mothers rated their children's well-being as significantly worse and reported significantly more somatic symptoms in their children, presumably due to Chernobyl-related stress. Bromet, E. J., D. Goldgaber, G. Carlson, N. Panina, E. Golovakha, S. F. Gluzman, T. Gilbert, D. Gluzman, S. Lyubsky, and J. E. Schwartz. 2000. Children's well-being 11 years after the Chernobyl catastrophe. *Archives of General Psychiatry* 57:563–571.

Figure 8.2
Counselors and volunteers help stressed and grief-stricken evacuees deal with the trauma of Hurricane Katrina in the Houston Astrodome, September 3, 2005. (Photo: Andrea Booher/FEMA.)

MITIGATING THE IMPACT

The manifestations of emotional distress are normal reactions to highly abnormal circumstances. Therefore, it is unrealistic to expect that the psychological and social impacts of a traumatic experience can be eliminated entirely. Through proper and timely intervention, however, the emotional distress can be significantly reduced and longer term effects such as posttraumatic stress disorder can be prevented. It may be difficult for responders to maintain a completely nonanxious presence while they do their work. However, through education and training, they can be less anxious and maintain their effectiveness while trying to help victims and co-workers. This section provides an overview of some of these early intervention strategies; interested readers should refer to the "Resources" section for pointers to more in-depth discussion and training on the subject.

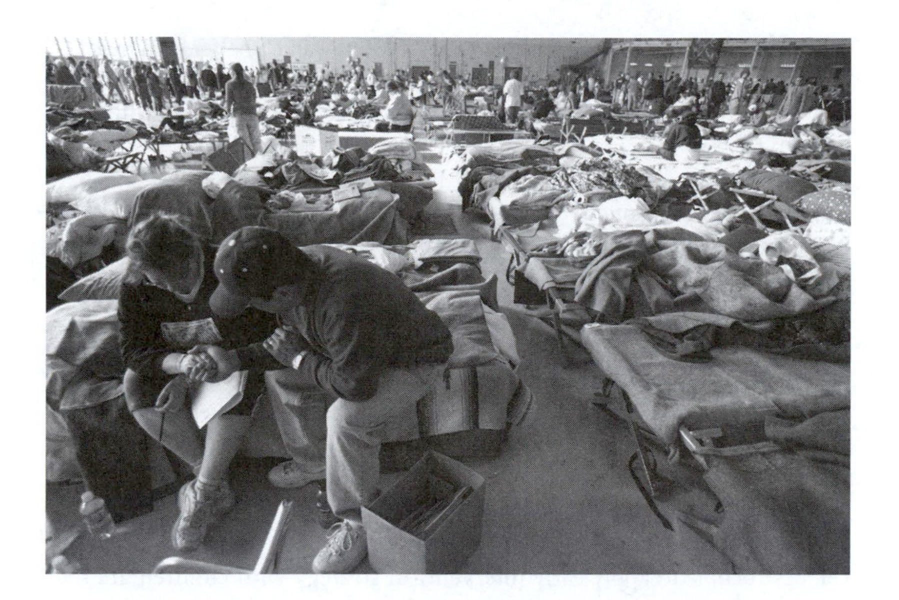

Figure 8.3
A volunteer clergyman prays with an evacuee in a shelter at Norton Air Force Base in San Bernardino, California, October 31, 2003. Hanger 3 at the air force base housed over 3,000 evacuees following the fires in Southern California. Note the Ronald McDonald clown (top right-hand corner) for the benefit of children at the shelter. (Photo: Andrea Booher/FEMA.)

Psychological First Aid

Psychological first aid (PFA) is an early intervention approach used to help disaster victims as well as emergency and disaster response workers in the immediate aftermath of a disaster. Mental health professionals or trained first responders can provide PFA at a shelter, medical facility, field triage area, or anywhere they are in contact with disaster victims or other response workers. However, it is important for any mental health assistance to be provided as part of a larger response effort. Treating mental health as a separate response unit or a separately labeled desk or counter at a shelter, for example, tends to stigmatize people, and victims are less inclined to go to that desk to seek assistance.

The ultimate goal of this early intervention is to promote, to the extent possible, an environment of safety, calm, and connectedness, and to instill a sense of empowerment and hope so that people are better able to help themselves. Some basic elements of PFA are to

- help people meet their basic needs of food, water, shelter, and medical attention in a safe environment;
- orient people to other support and services that may be available to them;
- facilitate communication with family and friends, and assist in locating loved ones;
- listen to those who want to share their stories, but do not force them to do so;
- offer accurate information if available;
- encourage rest, sleep, and normalization of daily routines to the extent possible; and
- give practical suggestions that steer people toward helping themselves and meeting their needs.

Children

The best people to apply early intervention strategy with children are their parents. Children often look to their parents to receive clues on how to feel and act and can easily pick up on fears and anxieties. Parents should try to maintain their composure. If they are overwhelmed with a sense of alarm or grief, children's reactions will be stronger too. Parents should also try to

- minimize children's exposure to frightening news and disaster images;
- spend time with children and show affection;
- allow children to express themselves and ask questions and answer their questions honestly, but not provide more details than they ask for;
- keep children busy by asking them to draw pictures or read stories;
- establish a routine to the extent possible. When normal routines are disrupted, children are unsettled. People who have been forced to evacuate their homes should try to establish new routines in the new location; and
- use humor when possible and appropriate.

Did you know? The 1997 award-wining Italian film *La vita è bella* (life is beautiful), depicts a Jewish Italian father (Roberto Benigni) in a Nazi concentration camp with his young son. During their entire period of captivity, he pretends to his son that they are engaged in some fancy hide-and-seek play with the German guards in an effort to help his son survive and to shield the child from the camp's atrocities. Although no mental health professional would recommend that particular coping strategy, the story demonstrates the importance of parent–child relationships as well as humor under such extreme circumstances.*

Disaster Response Workers

An essential element in preparing workers who respond to a radiation emergency—from front-line responders to medical and support staff—is prior training in basics of radiation and contamination control. This training must be appropriate to specific job functions and be augmented by additional just-in-time briefings. It also helps to have a support system in place to help staff communicate with their families and remain informed of their status and well-being during the response. As with any other disaster response situation, it is important to avoid exhaustion by limiting shifts and taking adequate breaks, working in teams, and having mental health professionals available to support staff when needed.

Education and Communication

People's resiliency in response to a radiation emergency is related to how much they know about the hazard and the quality of information they receive from the authorities. It may be unrealistic to plan for massive pre-event education campaigns and expect people to retain the information about a hazard they may not even regard as likely to affect them. The burden then falls on the quality of communication messages from the authorities during the disaster and the information people receive through the media.

Government authorities should use open and honest public communication strategies, using the best technical and scientific information available at the time and providing timely information people need to protect their health and safety. Even if long-term uncertainties remain, the ability to predict some future events with certainty helps to comfort the public. If an incident has not affected a group of people, but they are nevertheless concerned, the concerns need to be addressed, not dismissed. Losing trust in authorities will compound the extent of psychosocial impact that is already expected from such an incidence.

People are always advised to tune to radio and television to receive disaster-related information, preferably from credible emergency management

* Viktor Frankl, a psychiatrist and Nazi camp survivor, described the use of humor among camp prisoners as "another of the soul's weapons in the fight for self preservation." See page 43 in Frankl, V. E. 2006. *Man's search for meaning*. Boston: Beacon Press. (Originally published in 1946 as *Ein psycholog erlebt das konzentrationslager*; first English translation in 1959 under the title *From death-camp to existentialism*.)

and public health authorities. The role of the media cannot be overstated. Providing accurate information will serve the public and is one of the media's critical and difficult missions at times of crisis. People look for information about possible risks and what protective measures they should take. It is almost inevitable that experts will disagree, especially when all relevant information is not yet known. People should expect that such disagreements among TV "experts" are normal; unless people have good reasons to do otherwise, they should trust the information from responsible authorities.

Resources

The references in the footnotes, specifically the books by Robert Lifton, John Hersey, Gaynor Sekimori, and Peter Houts, are recommended for further reading. The following textbook covers a broad range of issues related to psychological trauma in general, with chapters addressing specific disasters such as floods, earthquakes, technological disasters, and terrorism:

Ursano, R. J., B. G. McCaughey, and C. S. Fullerton, eds. 1994. *Individual and community responses to trauma and disaster: The structure of human chaos.* Cambridge, England: Cambridge University Press.

The following two references provide an excellent discussion of psychosocial issues specifically in a radiation disaster:

Becker, S. 2001. Psychosocial effects of radiation accidents. In *Medical management of radiation accidents,* 2nd ed., ed. I. A. Gusev, A. K. Guskova, and F. A. Mettler. Boca Raton, FL: CRC Press.

National Council on Radiation Protection and Measurements (NCRP). 2001. *Management of terrorist events involving radioactive material.* NCRP report no. 138. Bethesda, MD. (See Section 5 in the report, which is on psychosocial effects of radiological terrorist incidents.)

For more information on surviving strategies from compassion fatigue and treating vicarious trauma, see:

Figley, C. R., ed. 1995. *Compassion fatigue: Secondary traumatic stress disorders in those who treat the traumatized.* London: Brunner-Routledge.

Rothschild, B. 2006. *Help for the helper: The psychophysiology of compassion fatigue and vicarious trauma.* New York: W. W. Norton.

For more information about psychological first aid, see National Child Traumatic Stress Network and the National Center for PTSD. 2006. *Psychological first aid: Field operations guide,* 2nd ed. (available from www.nctsn.org and www.ncptsd.va.gov).

Additional information and informative fact sheets are available from the Centers for Disease Control and Prevention (www.bt.cdc.gov/mentalhealth/) and the World Health Organization Web site on mental health in emergencies (www.who.int/mental_health/resources/emergencies/en/).

The Center for the Study of Traumatic Stress has a number of informative fact sheets on the topic of disaster and terrorism, including stress management for parents and intervention strategies (www.centerforthestudyoftraumaticstress.org/).

For an analysis of gaps in planning, preparedness, intervention strategies, and public health infrastructure in the United States, as well as experts' recommendations to address those gaps, see Institute of Medicine, National Academy of Sciences. 2003. *Preparing for the psychological consequences of terrorism: A public health strategy.* Washington, D.C.: National Academies Press.

Members of the media may be interested in the following reference guide specifically developed for them that includes useful information on a variety of terrorist threats. Appendices include a self-monitoring checklist for measuring stress levels and a guide for helping children and adolescents cope with disasters (the guide is free and can be ordered by calling 800-553-6847 or downloaded from www.hhs.gov/disasters/press/newsroom/mediaguide/index.html).

U.S. Department of Health and Human Services. 2005. Terrorism and other public health emergencies: A reference guide for media. Washington, D.C.

Part Three

RESPONDING TO RADIATION

9

PROTECTING YOURSELF AND YOUR FAMILY

OVERVIEW

THE CHAPTERS IN THE PREVIOUS SECTION reviewed how a radiation incident can affect health and environment. Some of these effects are negligible and should not be of concern. Others, however, can be serious and even fatal. Everyone needs to know what to do to protect themselves and their families and minimize the impact of potentially serious radiation hazards. In most emergencies, members of the public are in fact true "first responders" because they can take helpful, often critically important action before emergency response and health care professionals arrive to assist or communicate information.

The chapter starts by reviewing three fundamentally important principles of radiation protection. It then describes specific actions you can take, consistent with those principles, in four different radiation scenarios that involve (1) an unknown radioactive object, (2) a dirty bomb explosion, (3) a nuclear detonation, and (4) a nuclear power plant accident. This chapter can be read without having the information from earlier chapters. However, if you have read those chapters, you should find the information here to be common sense and, we hope, easier to remember afterward.

THREE FUNDAMENTAL PRINCIPLES FOR PROTECTION AGAINST RADIATION

You will learn in this brief section what all students of radiation and radioactivity learn in formal, 1-hour "radiation 101" awareness classes and in college-level radiation protection courses. They learn about the same three principles: time, distance, and shielding. These are important because they are the basis for every step you can take to protect your family and yourself from exposure to radiation hazards.

Principle 1: Time

Limit how long you spend near a radioactive source or stay in a contaminated area. Radiation doses are cumulative, so the longer you remain near a source of radiation, the higher is the dose you receive. If you need to enter a high-radiation area to perform a vital function (to save a life, shut off a gas valve, etc.), you should limit the time you spend there as tightly as possible—the shorter the time you stay there, the lower your cumulative radiation dose will be.

> *Did you know?* Professionals who routinely need to work in contaminated areas are required by law to have a "radiation work permit" prepared by a supervisor and approved by a safety officer. The permit documents and describes what a given worker needs to do while in the contaminated area and how long that work activity is estimated to take. Any radiation dose a worker receives while in the area is monitored closely and recorded in her or his personal safety record so that accumulated radiation doses over time do not exceed allowable limits.

Principle 2: Distance

Your exposure to radiation can drop dramatically if you simply increase your distance from a radioactive source or a contaminated object—that is, if you just move away from it. To illustrate, imagine that a radioactive pellet has dropped onto the floor just 1 m in front of you. If you double the distance—that is, if you move just one additional meter away, the radiation level will drop by a factor of four. If you move 10 m away, the radiation level will be 100 times lower!

The simple mathematical relationship at work here is that if you increase your distance from a radioactive source by a multiple of two, three, or more, the radiation level will drop by the square of that number. Regardless of which unit you use to measure distance (for example, meter, yard, or feet), when you double your distance, or increase it by twofold, radiation levels drop by a factor of 2^2, or fourfold. This relationship is called the "inverse-square" law and applies whenever you are dealing with a radioactive source that is physically small in size (i.e., a "point source"). The source could be very "hot" and present potentially lethal radiation levels. But as long as it is a relatively small object, the inverse-square law applies. In the example shown in Figure 9.1, a person standing within a 1 m distance of the unshielded radioactive source could receive a lethal dose of radiation within a few minutes to an hour.

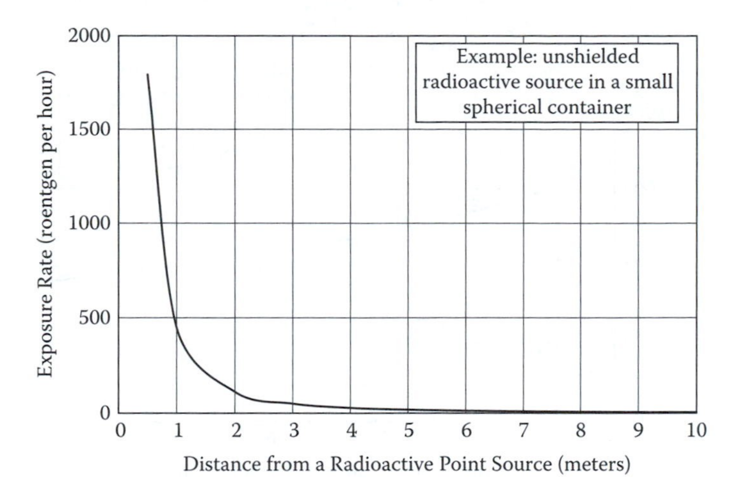

Figure 9.1

Radiation levels decrease dramatically with distance when a radioactive source is physically small in size (point source). The radioactive source used in this example is 50.9 TBq (1,375 Ci) of cesium-137 in a 30 cm^3 (1.8 inch^3) sphere—similar to the unshielded intact source from the Goiânia accident (see Chapter 4). A person standing within a 1 m distance of this radioactive source could receive a lethal dose of radiation within a few minutes to an hour. Stepping just a few meters (few yards) away from the source would take that person out of immediate danger.

Stepping just a few meters (few yards) away from the source would take that person out of immediate danger.

When you deal with a larger size source or area of contamination, however, the inverse-square law does not apply. In the example shown in Figure 9.2, the same radioactive source from the previous example is assumed to be uniformly spread over a 1 km^2 area (0.39 mile^2).* In this case, radiation levels still drop as you walk away from the contaminated area, but the drop is not as rapid as it is when you walk away from a smaller size radioactive source. Also, note the magnitude of radiation levels in both examples; when radioactivity is spread or dispersed, radiation levels are significantly lower.

In incidents when large amounts of radioactivity are spread over a large geographic area, if you cannot move away from the contaminated area quickly, you need to take other protective steps as described later in this chapter.

* If the 50.9 TBq (1,375 Ci) cesium-137 source from the Goiânia accident is uniformly spread over a 1 km^2 area, the concentration of radioactivity on the surface would be 5.1 kBq/cm^2 (0.14 μCi/cm^2).

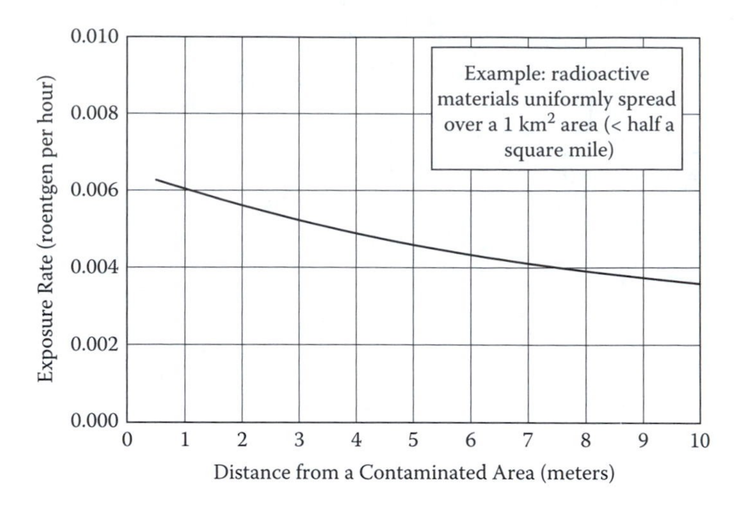

Figure 9.2

Radiation levels as we walk away from a 1 km^2 (0.39 mile^2) area contaminated with cesium-137. The total amount of radioactivity is identical to the example from Figure 9.1, but the radioactive material is uniformly spread over the area. Radiation levels do not drop as rapidly with distance because the source (contaminated area) is larger than a point source. In addition, note the magnitude of radiation levels in both examples; when radioactive materials are spread or dispersed, radiation levels are significantly lower.

Did you know? In 1974, an industrial-strength radioactive source of iridium-192 was shipped (legally) in the freight compartment of a Delta Airlines plane from Washington, D.C., to Atlanta, Georgia, and then on to Baton Rouge, Louisiana, where a radiation alarm went off. The protective shipping cask had malfunctioned and the radioactive source had shifted into an unsafe position, irradiating baggage handlers, passengers and crew on both flights. But, even though the radioactive source emitted high enough levels of radiation to inflict potentially serious harm, the radiation doses the exposed people received were low because the source was small in size and its radiation levels dropped rapidly with distance. It also was important that none of the baggage handlers had spent much time close to the exposed source.

Principle 3: Shielding

The intensity of radiation drops as it passes through certain types of barriers because radiation rays or particles interact with the atoms in the barrier and lose part or all of their energy. Therefore, standing behind such a barrier or placing certain materials between you and the source of radiation offers you protection. There are plenty of familiar examples. The dental technician places a lead apron on your chest when taking dental x-rays so that the rest of your body is not irradiated. Radiology technicians step behind a lead-lined wall when they take CT images of patients to avoid being irradiated. Fallout shelters provide shielding too. The shelter protects you from the penetrating radiation emitted by fallout particles that settle outside the shelter; if it is tightly sealed, the shelter also keeps radioactive particles from coming inside.

WHAT YOU SHOULD DO

Many people mistakenly believe that if they are exposed to radiation or if radioactive material gets on their skin, there is nothing they can do but be resigned to death. In fact, you can take simple steps in virtually all cases to reduce or eliminate your exposure to a radiation hazard. The following sections describe some of these actions in the context of four specific scenarios, making practical use of what you just learned about the principles of time, distance, and shielding.

Encountering a Potentially Radioactive Object

All radioactive materials or objects should carry a radiation warning symbol. Most of them do so most of the time. Some of the worst radiation accidents have occurred when a radioactive source is accidentally removed from its labeled and protective housing and is lost or misplaced at a construction site or other worksite. Such misplaced radioactive sources look like small, shiny pieces of metal and may appear innocuous and harmless. Tragic accidents have occurred when someone has picked up a radioactive source thinking it may have monetary value, placed it in a pocket, and taken it home. (See Chapter 4 for examples.)

Proper radiation detection instruments can easily determine whether an object is radioactive, but hardly any private citizens own or carry such instruments. As a child, you were taught never to "take candy from strangers"; similarly, you should never pick up and take away any unknown object you find, especially from a construction or industrial site. You should make sure that your children take the same precautions.

If you come across any object that is labeled with the standard radiation warning symbol (Figure 9.3) or other, similar marking, follow these instructions:

- Do not touch the object.
- Walk away from it.

- Warn others in the area.
- Contact the site supervisor or another manager, if one is available. You can also call your state radiation control agency (see "Resources" at the end of this chapter for contact information). If this does not work, call the National Response Center at 800-424-8802 and report your observations. The center operates 24 hours a day, 7 days a week.
- If you cannot walk away from the object at once or if help is slow to arrive, consider placing some kind of dense material—such as a piece of metal, or a bag of sand or soil—on top of the source until you can walk away from it or until help arrives. But, again, do not spend precious time trying to shield the object if you can just walk away.
- If you inadvertently touch the object, avoid touching your face or eating until you wash your hands with water (lukewarm, not hot) and mild soap if available. If you already have touched your face, wash your face with water and soap.

Figure 9.3

The international symbol for radiation and radioactivity is also known as trefoil. The most common color combination is black on yellow background, although magenta on yellow background is also used. Often additional wording such as "caution," "radioactive materials," or "radiation area" is added to the label. The symbol itself has no inherent meaning and only people who have been educated in its meaning have knowledge that it represents the presence of ionizing radiation.

These safety instructions should not cause any alarm. Coming in contact with a lost or displaced radioactive source is a rare event. Furthermore, most radioactive sources by far emit low levels of radiation and pose no hazard. Therefore, you do not need to worry that every unknown object is radioactive. But the prudent action is to refrain from touching suspicious objects and to keep children from picking up and playing with unfamiliar objects.

You probably have seen the familiar radiation warning symbol (Figure 9.3) on a hospital door or in a university research building. This generally means that a radioactive source or a radiation-generating machine is housed in the room behind that door. Standing outside the room or sitting in a nearby waiting room is perfectly safe. But you should not enter a room that bears a radioactive warning symbol unless an authorized person accompanies you.

In 2007, the International Atomic Energy Agency (IAEA) designed a new radiation warning symbol that uses radiating waves, skull and crossbones, and the figure of a running person (Figure 9.4). The purpose of this new symbol is to prevent death and injuries from accidental exposures specifically to radioactive sources containing large amounts of radioactivity, such as food irradiation machines, cancer therapy machines, and industrial radiography sources, rather than for sources with small amounts of radioactivity.* The new warning label will be placed on the housings of machines that contain these high-radiation-level radioactive sources to warn against dismantling the machines. Tragically, taking such devices apart has resulted in a number of radiation accidents.

Did you know? In the process of developing the new IAEA radiation warning symbol, the Gallup Institute field tested it with a total of 1,650 individuals in Brazil, Mexico, Morocco, Kenya, Saudi Arabia, China, India, Thailand, Poland, Ukraine, and the United States to make sure it conveyed the desired strong warning message to people who speak many different languages and who come from many different cultural and educational backgrounds.

Following a Dirty Bomb Explosion

The first thing to recall from Chapter 5 is that a dirty bomb explosion is not a nuclear detonation. A dirty bomb causes an "ordinary" explosion that disperses some amount of radioactive materials. The explosion is likely to cause a number of fatalities. For survivors, the radioactive material that the explosion disperses is a health risk and should be avoided, but the radioactivity

* Specifically, this new warning symbol is intended for IAEA category 1, 2, and 3 sources defined as dangerous sources capable of death or serious injury. See Table 4.1 in Chapter 4 for a description of radioactive source categories.

Figure 9.4

This new supplementary radiation warning symbol is intended only for radiation sources capable of causing death or injury as defined by the International Atomic Energy Agency (IAEA). This symbol was the result of a multiyear effort by the IAEA to develop a universal radiation warning sign so that anyone anywhere in the world would understand the message of "Danger—Stay Away." (Photo: © ISO. This material is reproduced from ISO 21482 with permission of the American National Standards Institute [ANSI] on behalf of the International Organization for Standardization [ISO]. No part of this material may be copied or reproduced in any form, electronic retrieval system, or otherwise, or made available on the Internet, a public network, by satellite, or otherwise without the prior written consent of the ANSI. Copies of this standard may be purchased from the ANSI, 25 West 43rd Street, New York, NY 10036, 212.642.4900 [http://webstore.ansi.org].)

is unlikely to be fatal, even close to the site of the explosion. In all likelihood, people on the scene will not know immediately after the explosion that radioactive materials were involved. The professional first responders who arrive on the scene and certainly those on duty in most large metropolitan areas have the equipment necessary to make that determination shortly after they arrive.

At or Near the Site of the Explosion If you are outdoors when a dirty bomb explodes (assuming you know what it is), you should immediately cover your nose and mouth with a piece of cloth or handkerchief and quickly go inside an *undamaged* building nearby. Most people are not trained engineers who can technically evaluate the extent of damage to a building. For our purposes, undamaged simply means any building with walls, doors, and windows that are structurally sound that is not at risk of collapsing on top of you. If you cannot identify such a building nearby, move quickly away from the explosion on foot in an upwind direction until you find such a building and go inside it. If you are driving when the explosion occurs, close the windows, park the car, and go inside a nearby undamaged building. If you must stay in the car, keep the windows closed and turn off the air conditioner.

If you are inside a building or at home when you hear the explosion, stay where you are. Emergency management professionals call this "sheltering in place." People can shelter in place in a school, an office building, a hotel, a restaurant, or their own home if that is where they are when the incident occurs.

Once inside the building, follow these instructions:

- Close doors, windows, vents, and fireplace dampers and turn off the heating or air conditioning and anything else that could bring outside air into the building. A word of caution: Be sure that there is enough air to breath and that inside temperatures do not get too high. This is especially important if children or elderly people are present in the building. Make sure that breathing air and temperatures are not compromised even if you have to turn the air conditioner on or open a window. Dealing with a minor "nuisance" amount of radioactivity is a better alternative than dying from heat or suffocation because you sealed your shelter too tightly.
- Eat and drink only food and water that has been covered. Do not eat or drink anything that may have been uncovered or left in the open where radioactive dust from the explosion could have settled on it.
- Use the radio, television, or Internet to get updated health and safety announcements. The authorities will inform you when it is safe to go outside or, if necessary, to evacuate to another location. As is described later, battery-powered radios are indispensable in situations when loss of electrical power is a possibility.
- In the case of a dirty bomb, the period of sheltering will be short, extending until the plume of dust from the explosion has passed (30 minutes to a few hours, depending on location and weather conditions). At that point, official announcements will tell you it is okay to open windows for fresh air. The radioactivity levels on the ground outside are not likely to be high or life threatening. It will be safe to walk outside to perform essential duties, although the possibility of getting contaminated with radioactive materials would still be present.
- Whenever you feel you may have been contaminated with radioactive materials, follow the simple decontamination steps presented in the next section.

- If you must go outside when dust from the explosion is still visible in the air, use a piece of cloth or a handkerchief to cover your nose and mouth. Your shirt tail would work, if need be. Even a piece of paper towel or folded sheets of toilet paper can offer some protection against inhaling radioactive particles.*
- If you are helping children or elderly people, make sure that whatever is used for respiratory protection does not restrict their breathing. Again, dealing with some amounts of inhaled radioactivity is a better alternative to dying of suffocation.

Some Distance Away from the Explosion The sound of a powerful explosion, like that of a dirty bomb, may be heard miles away. Remember that the immediate health risk from a dirty bomb is the explosion itself. Therefore, if the building you are in is undamaged, stay there and follow the preceding instructions. If a large radioactive source is used in the dirty bomb, authorities may advise people within a 1- or 2-mile radius of the blast to stay where they are until the plume of dust has passed.

After the plume has passed, depending on the situation on the ground, authorities may advise people in the immediate area to evacuate. This area is likely to be limited to a radius of 1 mile or less from the site of the explosion. Until authorities learn more about the situation and make appropriate statements, you should stay away from the immediate area of the blast. Even though it goes against your parental instincts, it is wiser to let your children stay sheltered in their schools until the authorities give the all-clear signal, probably in just a few hours.

If you are dozens of miles away from the explosion, you should be minimally affected, if at all. Nevertheless, you should stay up-to-date on official announcements and guidance by listening to authoritative sources of information. It is possible that a much larger area—on the order of 100 $mile^2$ or more—will receive some small, residual amounts of radioactivity. People in that area may be advised to relocate temporarily until the area is cleaned.

It is important to understand that an official advisory notice to relocate does not mean that everyone should rush onto the highways. Too

---

* Using such devices for protection is far from ideal because they do not form a tight seal around your nose and mouth. Therefore, using these materials in an emergency should not create a false sense of security. See Guyton, H. G., H. M. Decker, and G. T. Anton. 1959. Emergency respiratory protection against radiological and biological aerosols. *AMA Archives of Industrial Health* 20 (2):91–95; Cooper, D. W., W. C. Hinds, and J. M. Price. 1983. Emergency respiratory protection with common materials. *American Industrial Hygiene Association Journal* 44 (1): 1–6; and Cooper, D. W., W. C. Hinds, J. M. Price, R. Weker, and H. S. Yee. 1983. Common materials for emergency respiratory protection: Leakage tests with a manikin. *American Industrial Hygiene Association Journal* 44 (10): 720–726.

hasty evacuations can lead to congestion, crashes, and related injuries and deaths. Emergency management officials issue relocation advisories out of an abundance of caution and based on radiation doses projected over 1 or 2 years into the future, meaning that there is plenty of time for an orderly relocation. It is important to stay tuned for information from the authorities. Keep in mind that the rumor mill is active during emergencies, both among your neighbors and in the media. Do not make any decisions based on rumors.

> *Terminology:* There is an important difference between *evacuation* and *relocation* advisories. Evacuation is a more urgent action for the purpose of protecting people from a short-term exposure to what could be high levels of radiation. Relocation, on the other hand, is not an urgent action. If issued, relocation advisories are for the purpose of protecting people from low levels of radiation exposure projected over a 1- or 2-year period, or even longer.

Instructions for Self-Decontamination at Home

Contrary to what many people may believe, dealing with radioactive contamination on your clothes or body is fairly simple. If you think you have been contaminated, follow these simple steps:

- Remove the outer layer of clothing (your coat or jacket), place it inside a plastic bag, and tie the bag shut. You can carry this bag home later and place it in a corner of your garage or in your car trunk. Keep the bag there until the authorities advise what to do with it. Your bag of clothes can help health professionals determine whether you were contaminated and, if so, to what degree. Leaving the bag of clothes in a garage corner or in your car's trunk will not harm you or anyone else.
- When you arrive at home (or another destination), remove your shoes. Remove the rest of your clothes and place the shoes and clothes in a bag and tie the bag shut. If possible, do this outside your home (for example, in a garage) to minimize tracking the material inside your home. When you are tying the bag shut, do not squeeze the bag to let the air out because this may blow out some of the dust.
- When you remove your clothes, do not take them off over your head. If, for some reason, you have to remove a shirt, sweater, or other clothes over your head, hold your breath while you take them off. This will keep you from inhaling any dust that may shake off your clothes.

- If you have any cuts or abrasions on your body, cover them while removing your clothes to prevent any contamination from radioactive dust.
- Go inside and take a shower using lukewarm water. Do not use hot water and do not scrub your skin with abrasive sponges or brushes. Hot water and strong scrubbing may increase the likelihood of radioactive material being absorbed through your skin.
- In the shower, start with your head and pay particular attention to your hair. Use a shampoo without hair conditioner. Conditioners may bind radioactive material to your hair proteins, making it more difficult to clean.
- Wash from the top down. Use a gentle detergent or soap. If soap is not available, just use water.
- Pay attention to your ears and eyes and rub them and your face clean.
- You do not need to collect the wash water. Let it drain as it normally does.
- If you do not have access to a shower, use a sink and wash as best you can. If you do not have access to running water, use moist wipes to clean your hands and face or use a small amount of water to wet a paper towel and then use it to clean your hands and face. Pay particular attention to the areas around your mouth, nostrils, and eyes.
- If you have infants or young children, hold them as needed while you wash and clean them as you normally would. However, avoid immersing them in water; you do not want your infant to bathe in potentially contaminated water. Showering is best. You can also wash an infant in a sink if that is easier.

An Easy Way to Remember How to Decontaminate Rather than think about radioactivity as such, imagine that you and your clothes are covered in ordinary mud. Imagine that you have no opportunity to clean up and that you need to get into your car and drive home. All the actions you take to protect your car from getting dirty and to avoid tracking the mud inside your house are exactly the right actions for you to take if you are contaminated with radioactive materials.

First, put your muddy coat or shirt inside a plastic bag, if you have one, and put the bag in the trunk of your car. Next, place a newspaper or some other similar material on your car seat to protect it from getting muddy. If your hands are muddy, you will probably try to avoid touching things that you do not have to touch. You can wipe the steering wheel and gear shift clean later.

Once you get home, you will remove your muddy clothes outside. If your clothes are very dusty, you should be careful not to inhale as you take your clothes off. These common-sense steps are the very same steps you should take if you ever are contaminated with radioactive materials. Therefore, just pretend you are covered in dust and mud and you will take all the right actions.

What If I Do Not Clean Myself Completely? Taking off your outer layer of clothing and washing yourself as described earlier should remove all the contamination on your body. If any remains, it will be very slight and inconsequential. If you leave a small amount of radioactive material in your hair, track a small amount inside your home, or inhale a small amount, there is no need to worry. Your health will not be in danger. Later (this could be several days after a dirty bomb explosion or other release of radioactive material), local health authorities will set up centers where you and your family can go to get checked. You may be asked to provide urine samples. Also, radiation professionals may be sent to check your home to check for any residual contamination, if necessary.

What If I Touch Someone before I Am Clean? When you are contaminated, the best practice is, of course, not to touch anything you do not have to touch until you have cleaned your hands. The most important precaution is to avoid drinking, eating, and smoking when radioactive contamination is on your hands or face. If you touch anyone, you may cross-contaminate him or her. That is, you may transfer a small amount of radioactivity to the other person. But, as you learned from previous chapters, this does not mean that the person will contract a contagious disease from which he or she will suffer terribly.

Think dust and mud again. When you touch someone with your dusty or muddy hands, he probably will get a little dirt on himself also. This just means that he must clean himself. If your young children are with you when you get contaminated, chances are that they will be contaminated as well. Comfort them, if you need to, by holding and soothing them. Ask them not to put their fingers in their mouths or eat any snacks in their pockets until you can get home or have an opportunity to wash their hands and faces.

What If I Cannot Get Home Quickly Enough? Even if it takes you several hours to get home after a radiation emergency has taken place, there is no cause for alarm. (The situation would be very different in the case of a nuclear detonation, when you should avoid being outdoors and exposed to nuclear fallout. This will be discussed in the next section.) Following a dirty bomb explosion, taking off your outer layer of clothing and putting it in a bag or car trunk should remove most of the contamination. Any radiation exposure from remaining radioactive material is most likely minimal. Wash your hands and face or rinse them clean at the first opportunity.

> *Did you know?* The 1983 Oscar-nominated film *Silkwood* contained a scene where a female worker (played by actress Meryl Streep) in a plutonium processing facility was scrubbed harshly in the shower to remove radioactive contamination from her body. As you learned in this section, such harsh measures are completely unnecessary.

Following a Nuclear Detonation

As discussed in Chapter 5, a nuclear detonation would be far more catastrophic than a dirty bomb explosion; a very large number of people would perish. (If you read Chapter 5 quickly, it may be helpful to read the section on improvised nuclear devices and become familiar with the effects of a nuclear explosion, the initial flash and intense heat, the shockwave that follows, and subsequent nuclear fallout.) After the initial shockwave or blast, survivors can take important protective steps to increase their likelihood of survival. In principle, these protective actions are the same as in a dirty bomb incident, although there are some important differences because

- you may be physically injured or burnt by the blast and heat;
- radiation levels associated with fallout are dangerous and life threatening;
- the affected area would be much larger;
- finding a suitable shelter would be more difficult;
- the sheltering period would be longer;
- it would take longer for help to arrive; and
- evacuations would take much longer.

What to Do Immediately after a Nuclear Blast The following discussion assumes that you receive no warning before the nuclear blast. This would almost certainly be the case in a nuclear terrorism incident. If there is advance warning, the best possible course of action is to take shelter in a strong structure for protection from the heat and blast of the incoming explosion (see Chapter 5). Large steel or reinforced-concrete buildings and basements are best; single-story frame houses are not as good and shed-type industrial buildings are worst.

At the instant of detonation, there will be a very bright flash of light. Follow these recommendations:

- Never look at the flash or fireball; it can blind you from miles away and, if you are close enough, it can literally burn your eyes. The intense light of the nuclear flash is even more hazardous if the detonation occurs at night.
- Take immediate cover behind any solid barrier that might offer protection—a wall or under a desk, but away from windows. Lie flat on

the ground if you can and cover your head. A ditch or culvert will provide good cover. Taking this basic step can significantly enhance your chance of surviving the initial effects of a nuclear blast. It will spare you from some of the initial radiation that will persist for 10 seconds or more. It will also protect you from the violent shockwave that follows shortly thereafter.

- Stay in that sheltered position for at least 2 minutes. Depending on your distance from ground zero, it could take several seconds or more for the blast wave to reach you. Once it hits, the shockwave should last only a few seconds. There could be flying debris, windows shattering, and objects thrown at high speed. You need to stay down until the shockwave has come and gone.
- After the shockwave has passed, gas line ruptures, secondary explosions, and multiple fires are likely in the heavily damaged area, and highly radioactive nuclear fallout will soon follow. At this point, it is important to avoid the physical dangers posed by fire, explosions, and damaged buildings, and protect against radiation levels, which will be extremely high and life threatening within a one-mile to two-mile radius of ground zero.
- If possible, seek shelter inside a safe building as soon as you can.

The most important function a shelter plays shortly after a nuclear detonation is to protect its inhabitants from fallout particles. The most hazardous fallout particles will be readily visible as fine, sand-sized grains that start to settle within a few minutes of the detonation.* In the initial hours, fallout will be highly radioactive and life threatening. It is vital to avoid exposure to nuclear fallout in the first few hours and days following detonation. If fallout particles settle on you or your clothes, gently brush them off. If you must go outdoors, first cover yourself with anything you can find. Use a hat or coat to cover your head or use a sheet of plastic or paper to hold over your head and keep fallout from settling directly on your body and clothes. Afterward, throw that cover away or use it again after gently dusting it off. Always shake things off gently to avoid inhaling radioactive dust.

Deciding Whether to Seek Shelter or to Evacuate The two principal actions for protection against nuclear fallout are to take shelter or evacuate. Choosing between these two immediate actions is one of the most difficult decisions for

* Smaller fallout particles may not be visible. Therefore, lack of visible fallout at distances away from the blast does not necessarily mean there is no radiation in that area. With proper instrumentation, response authorities can determine quickly whether fallout radiation is present and the radiation levels present a hazard.

survivors of a nuclear blast and also for emergency responders and authorities evaluating the situation as it changes and trying to provide the best recommendation possible under the circumstances. Let us consider some of the factors that should influence the decision; this may help clarify the best option:

- In the most heavily damaged area, there may be no safe place to shelter from radiation. Radiation levels will be immediately life threatening, there will be multiple fires and secondary explosions, and any standing buildings may not be structurally safe.
- Roadways will likely be clogged with abandoned, wrecked or overturned vehicles, and other debris. Consider the usual Friday afternoon as suburban commuters leave downtown in good weather and without any accidents. Then imagine, in contrast, what traffic might be like with masses of scared, injured, and disoriented people trying to flee after a nuclear detonation, using roadways clogged with debris and abandoned vehicles.
- The textbook path for evacuation to evade nuclear fallout is the "lateral evacuation route." What this means is that people should move in a direction perpendicular to the direction of incoming fallout. For example, if the fallout cloud is moving from west to east, it is best to evacuate to the north or south. The problem, however, is that actual fallout patterns do not necessarily follow textbook examples. The direction of the wind may change. Furthermore, winds may move in opposite directions at different altitudes at any given time and lead to unpredictable fallout patterns. No one can precisely predict the direction of fallout in the immediate hours after the explosion.
- Being stuck in your car would provide little protection against penetrating fallout radiation—roughly 10% of the protection you would have in a basement.
- Trying to evacuate the area may actually put you at a greater harm if you inadvertently move through a highly radioactive fallout area while evacuating.

In light of these factors, unless you are in a heavily damaged area near the blast site, the best action is to seek shelter immediately. Radiation levels 48 hours after detonation drop to only 1% of what they were 1 hour after detonation. In 2 weeks, radiation levels drop 10 times further to 0.1% of the original level. It may be best to stay in a shelter for 1 to 3 days and then attempt evacuation. Once inside a shelter, use a radio to get official guidance from the emergency response authorities, including when it may be advisable to evacuate.

Did you know? On March 1, 1954, the United States tested a hydrogen bomb (code-named Bravo) at Bikini Atoll in the Pacific Ocean's Marshall Islands. Inhabitants of nearby islands were not warned because meteorologists had predicted the fallout pattern would not affect them. Tragically, the wind shifted and the plume moved in a different direction. As a result, the inhabitants of Rongelap Atoll, 150 miles away, were exposed to high fallout radiation doses. During the same test, Japanese fishermen onboard the fishing vessel *Lucky Dragon,* 80 miles away, also were unwittingly exposed to what was later called "ashes of death." All the fishermen and many islanders developed radiation sickness. One of the fishermen died from exposure to nuclear fallout.

Where to Shelter Sheltering in an underground basement is an excellent option (Table 9.1). If your house has a walkout basement, the side surrounded by earth and away from windows is the better side to use for shelter. If you are in a multiple-story or high-rise building, staying on one of the middle floors and away from windows offers better protection than the first floor or the top floor. The first floor is not as suitable because you are closer to the fallout that settles at street level. The top floor of a flat-roofed building is not a good option because flat roofs collect fallout particles; on the top floor, you will be closer to the radiation that they emit. The more you can have heavy and thick items or barriers between you and the fallout particles, the better your protection will be. Subways, underground tunnels, and underground parking garages can also be used as shelter.

Single-story frame homes without basements are the least desirable place to shelter. If you are in such a building, try to shelter in the middle of the house, away from exterior walls and windows. If heavy items like bricks, cinderblocks, or even boxes of books are available, you can use them as additional barriers.

What to Do inside the Shelter If a significant amount of fallout dust is on you or your clothing when you reach the shelter, remove your outer clothing and brush it off before entering the shelter. But do not delay entering the shelter in order to remove the last speck of fallout dust. Tracking a small amount of fallout inside the shelter is not important at that point.

Once inside, try to secure doors, vents, windows, and other openings where wind may blow fallout particles inside the shelter. But be mindful that you and other occupants need ample ventilation to breath. Turn the heating or air conditioning system off unless it is essential to the supply of air. Contamination of food and water supplies is a concern too. Any food and water stored in the shelter earlier can be used. Food and water brought from outside should be checked for contamination. If any dust is visible on food and water containers, they should be wiped or cleaned. Fruits and vegetables

Table 9.1 Structures' Protective Effects against Nuclear Fallout

Structure or Location	Protection Factor[a]	Comments
Outside	1	A person standing against a building would be exposed to less radiation than a person standing in the intersection of two streets or out in open areas away from buildings.
Vehicles (cars, buses)	1–2	Vehicles provide minimal shelter.
Frame house	2–3	Single-story wood frame homes without basements provide minimal shelter. Brick homes are better.
Basement (with partially exposed walls and windows)	2–10	The corner of a basement that is under earth and not exposed offers better protection. Use any dense material available to block the exposed side of the basement.
Basement (without exposed walls, under earth)	10–50	The basement of a two-story home offers better protection than that of a single-story home.
Multiple-story building (e.g., office or apartment)		Best protection is available on a midlevel story, middle of the floor, and away from exterior walls. The top floor is not ideal because of proximity to the roof, where fallout accumulates.
Upper stories	100	
Lower stories	10	
Underground shelter, sub-basement of multiple-story building	5,000	Assume at least 1 m (3 ft) of earth cover or the equivalent.

Source: Adapted from Glasstone, S., and P. J. Dolan. 1977. *The effects of nuclear weapons,* 3rd ed. Washington, D.C.: United States Department of Defense; and Homeland Security Council. 2009. *Planning guidance for response to a nuclear detonation.* Washington, D.C.

[a] The protection is against gamma radiation from nuclear fallout. The protection factor for each structure is relative to that of standing outdoors with no shelter. The numeric values are only approximations because exact protection factors depend on a number of variables, such as distribution of fallout depositing on the structure, adjacent buildings or trees, and the specific construction for each structure.

suspected of being contaminated should be washed with water or peeled before eating. Stay tuned to the radio or TV for information, but keep in mind that broadcasts may be disrupted for some time because of power outages.

When to Leave the Shelter As time passes, radiation levels associated with nuclear fallout drop rapidly. A well-established rule of thumb is that for every sevenfold increase in time past the detonation, radiation levels drop 10-fold. Based on this 7/10 rule, the radiation dose rate 48 hours after detonation is only 1% of the dose rate 1 hour after detonation. After 2 weeks, radiation levels drop to 1/1000 (0.1%) of what they were 1 hour after the blast.*

Therefore, the most critical time to shelter in the 30-kilometer (20-mile) downwind distance of ground zero is the first hour after detonation, when most of the fallout settles on the ground and is most highly radioactive. Sheltering, even if it is just for that first few hours before evacuating, is effective in reducing your total radiation exposure (see Figure 5.6 in Chapter 5). But you likely will need to shelter longer than that (at least for several days and perhaps for a couple of weeks), assuming that you and family members have access to food, water, and any essential medications or other supplies.

The further you are from the detonation site, the more time you will have to seek shelter before the cloud of fallout arrives and the more time you have to stock a shelter with supplies. The fallout that settles on the ground at distant locations may not be visible, but it can be easily detected by proper instruments. Fallout particles that settle far from the detonation site decay considerably during their time in the atmosphere. This means that radiation levels will be much lower compared with the levels near ground zero. In time, you will be able to leave the fallout shelter. Stay tuned to the radio or TV, if available, to receive official information and instructions from authorities.

Following a Nuclear Reactor Accident

Although it is serious, an accidental radiation release from a nuclear reactor would be far less dangerous than a nuclear explosion. The characteristics of such an incident that account for this dramatic contrast include the following (see Chapter 4 for more details and explanation):

- It is impossible for a nuclear detonation (like a nuclear bomb) to occur as a result of an accident at a nuclear power plant.†

* As stated in Chapter 5, these values are approximations. The main point is that radiation levels decrease rapidly in the hours and days immediately following a nuclear detonation.

† The type of nuclear fuel used in a power reactor, physical dimensions of that fuel, and how it is packaged make it impossible for the reactor to detonate like a nuclear bomb.

- Accidents at nuclear power plants generally follow a predictable chain of events; there will most likely be advanced warning to the public before any radioactivity is released to the outside as a result of the accident (see Chapter 4).
- There will be adequate time to seek shelter or evacuate if recommended.
- You will not be burned or injured.
- Radioactivity levels released from the plant may be high, but they will not be immediately life threatening, as in the case of nuclear fallout.
- Communication systems to provide information to the public will most likely be working.
- In the worst-case scenario, residents living within a certain radius of the nuclear power plant may be asked to evacuate and may not be able to return to their homes for some time. It is highly unlikely that anyone outside the plant would die from the accident unless in a traffic-related accident caused by hasty evacuation.

Regarding the radioactive materials released from the reactor, any deposits of radioactive particles on your clothes or body will not be visible—unlike nuclear fallout, which consists of visible clumps of dust. If you think you may be contaminated following a release of radioactivity, you should follow the exact same steps described earlier for decontamination (i.e., changing your outer layer of clothing and showering as soon as you can). The urgency to avoid this radioactivity is not the same as it would be with nuclear fallout. Even if you remain contaminated for the initial several hours, you will not receive a life-threatening dose from that radioactivity. Sheltering in a single-story home is fine as long as you are able to close doors, windows, and vents during the short sheltering period. You do not necessarily have to move to a basement or the middle floor of a multiple-story building.

Therefore, power plant accidents, should they ever occur, are far more manageable scenarios for the public and emergency responders compared to a nuclear detonation incident. Keep in mind that even in the absence of any significant releases of radioactivity, an accident at a nuclear power plant will cause anxiety and stress (Chapter 8).

But, you may ask, what about Chernobyl? Did that 1986 accident not prove that a nuclear power plant accident can be devastating? Absolutely. But it is important to appreciate that the consequences of the Chernobyl accident resulted directly from the poor safety features of the plant, poor safety procedures, and the failure of Soviet authorities to acknowledge the accident and inform citizens for several days after the accident (Chapter

4). There is no reason to believe any circumstances remotely similar to Chernobyl could occur in the United States.*

If you live within a 10-mile radius of a nuclear power plant (referred to as the 10-mile emergency planning zone [EPZ]), you should be familiar with emergency response procedures and alert systems in your community. If you are not, contact your state's radiation control program and ask for that information (see "Resources"). If you ever receive a warning of a potential or actual accident, tune your radio or television to the emergency alert system station for your area. This station should be identified in the emergency preparedness information you receive annually from your county government as a resident in the 10-mile EPZ. Follow the instructions you receive from this station. Your instructions may include directions for evacuating or for remaining in place, which is the same as the "sheltering in place" discussed earlier.

If officially recommended, sheltering following a nuclear power plant release will be for only a short period of time—far shorter than after a nuclear detonation. Recommendations also may be issued to use only food and water from closed containers. In affected farming and rural areas, the officials may recommend removing animals, especially dairy animals, from pastures; sheltering them, if possible; and providing them with stored feed and water. If any animal feed is uncovered, it should be protected with a tarp. Similarly, any open sources of water (such as cisterns, wells, and livestock watering tanks) should be covered.

In all EPZ communities surrounding nuclear power plants in the United States, such accident scenarios are played out and practiced regularly. Emergency responders and managers practice what they need to do to protect the public in the unlikely event of an accidental radiation release from a power plant. Appropriate instructions are ready for public communication as needed. Schools and hospitals in those areas also have specific plans on how to respond to such an emergency. You should be familiar with the plans, take part in community meetings that are convened from time to time to receive public input on them, and, in the event of an actual radiation release, always be alert to official announcements and carefully follow guidance from local, state, and federal agencies.

---

* One major difference is that the Chernobyl reactor used graphite as moderator, which made it vulnerable to loss of water and extreme temperatures. When water is lost, chain reactions accelerate. In contrast, U.S. commercial power reactors use water as moderator, which also serves to self-regulate against loss of water and overheating (i.e., when water is lost, chain reactions stop). Another difference was the containment structure. The Chernobyl containment design could never be licensed in the United States. Safety practices and procedures are also more restrictive in the United States.

SPECIAL CONSIDERATIONS

Almost every family has members and friends who may have special needs or require special consideration in any type of an emergency including a radiation emergency. This section discusses some of these issues and recommends steps you may consider taking to safeguard their well-being during such an event. Many families also want to know how to protect their pets and prevent possible radioactive cross-contamination from pets to children and adults.

Children

You should be aware of children's emotional needs during any type of an emergency. They may become confused, anxious, or frightened. Because they will not fully understand what is going on, children will be looking to their parents to figure out how concerned they need to be. You should try to explain what is going on and reassure them. You can do this best by staying calm. The American Academy of Pediatrics has a number of resources related to children's needs in disasters (www.aap.org/disasters). See Chapter 8 for more discussion on this and related topics.

When children are in school, day care, or similar settings out of the home, many parents' first instinct when an emergency occurs is to rush to that location and bring their children home. If emergency instructions are to take immediate shelter, the safest place for children is at the school where the adults in charge can shelter them appropriately. Schools have emergency response plans in place for emergencies related to severe weather and natural disasters. The same sheltering procedures can be used in a terrorism incident such as a radiation emergency. If school, day care, and other administrators or staff have not brought these issues to your attention, you will want to take the initiative to learn about them and about the steps you should take in an emergency.

Nursing Mothers and Pregnant Women

Under most likely circumstances, nursing mothers can continue to feed their babies safely during and following radiation emergencies. Let us examine situations in which a mother is irradiated, externally contaminated, or has internal contamination—for example, from inhaling or ingesting radioactive dust or particles:

- If a mother is irradiated (exposed to radiation), the mother's milk is safe. Nursing infants will not swallow any harmful substance with the milk even if the mother has been exposed to large doses of radiation.

- If there is a chance that either the infant's skin or the mother's skin has been contaminated with radioactive materials, the infant's face and hands as well as the mother's breast and hands should be cleaned before feeding to minimize chances that the infant might swallow any radioactive dust or particles. If possible, the mother should wash herself and the baby with water and gentle soap before feeding the baby.
- If a mother has some internal contamination or is at risk of becoming internally contaminated, some fraction of that radioactivity is likely to be transferred to her milk. Consequently, it is best if she stops breast-feeding and switches to using baby formula or other food until she can be tested by a health professional. But if breast milk is the only food available for an infant, nursing should continue. Even if some amount of radioactivity is transferred to the baby through the mother's milk, the health consequences would be far less significant than dehydration or lack of food.

Pregnant women should know that their unborn children are better protected inside the womb than they would be outside it. Any steps a pregnant woman takes to protect herself from radiation also protect her unborn child. Of course, pregnant women should be given high priority if food and water, shelter, showers, and other critical services and supplies are limited in the period following a radiation emergency.

Other Vulnerable Groups

Several groups of people are especially vulnerable in any type of an emergency. These include some of the elderly, people with physical or mental disabilities, people who may not understand or speak English, the hospitalized population (especially those in intensive care), and people in nursing homes, group homes, jails, and other institutions. Elderly family members, for example, may have reduced mobility, may have decreased tolerance for extreme heat or cold, and may need medications on a regular basis. Families and emergency responders need to identify these individuals and groups and make plans for their protection in an emergency. Many communities have special emergency plans in place to support and protect people who are especially vulnerable.

Pets

Many pet owners fear for their pets' well-being in emergency settings. (People who rely on seeing-eye dogs and other service animals have great concern as well.) This was made very clear when Hurricane Katrina ravaged the Gulf Coast in 2005. That dramatic experience led Congress to enact the Pet Evacuation and Transportation Standards (PETS) Act of 2006.* The PETS Act requires that state and local emergency plans address the needs of people who have household pets and service animals. Consistent with that legislation, the federal government now recommends to pet owners: "If you evacuate your home, do not leave your pets behind! Pets most likely cannot survive on their own, and if by some remote chance they do, you may not be able to find them when you return."†

When family emergency plans are discussed in the next chapter, the discussion will also describe how to plan for pets and service animals in a radiation emergency. In response to the emergencies discussed earlier in this chapter, whatever actions you take to protect yourself also will be protective of your pets. However, you do need to take some extra precautions:

- If you shelter inside your home, bring your pets inside with you. Do not leave them outside.
- Decontaminating your pet may present more of a challenge than cleaning yourself or your children. There is no outer layer of clothing to take off. Pets with long hair may be harder to decontaminate.
- Keep in mind that a level of contamination that does not harm you will not harm your pet. But be aware that your children may play with pets and then proceed to put their hands in their mouths or eat without washing their hands. If possible, keep a pet that you know is contaminated with radioactive materials, or suspect it to be contaminated, in the garage or some other location where it will not interact with family members.
- If you bring your pet inside your home, make sure the doors are closed (or locked) so that the pet does not wander outside, where it may pick up radioactive contamination and bring it back inside.
- Have newspapers on hand for sanitary purposes. Feed your pets moist or canned food so that they will need less water to drink.
- Pets are sensitive to humans' emotions, including the anxiety everyone will feel following a radiation emergency. Pets may react with unusual and even troublesome behavior. Be alert to their behavior and take steps to calm and reassure them.
- If you must evacuate, take your pets too.

---

* Public Law 109–308, October 6, 2006.

† Federal Emergency Management Agency Web site (www.fema.gov/plan/prepare/animals.shtm). Accessed March 20, 2009.

A FINAL WORD

There is no reason to feel helpless when faced with a radiation emergency. You can take certain actions on your own to protect yourself and your family. Even in a devastating nuclear disaster, specific actions can be taken to increase the likelihood of survival. Remember that any protective step is better than none at all. Because preparedness is a key component of resilience, the remaining chapters will discuss family emergency plans and how to prepare for any radiation emergency ahead of time as families, as professionals, and as communities.

Resources

To find the radiation control authority in your home state, visit the Conference of Radiation Control Program Directors (CRCPD) Web site (www.crcpd.org).

For farmers, food producers, and distributors, specific instructions and recommendations are provided by each state's departments of health, agriculture, or emergency management. The following are a few examples:

- Washington: http://agr.wa.gov/foodsecurity/attachments/radiological%20emergency%20book.pdf
- Virginia: http://www.vdem.state.va.us/threats/radiological/
- New Jersey: http://ready.nj.gov/farmers.html

Several authoritative fact sheets are available from the following two Web sites on actions to take in response to nuclear or radiological emergencies:

- Centers for Disease Control and Prevention: emergency instructions for individuals and families (http://emergency.cdc.gov/radiation/emergencyinstructions.asp)
- U.S. National Academy of Engineering fact sheets on terrorist attacks (www.nae.edu/nae/pubundcom.nsf/weblinks/CGOZ-642P3W?OpenDocument)

The following textbook provides a wealth of easy-to-read information on an extensive range of terrorist threats in a consistent format and includes many pages of information on radiological and nuclear emergencies:

- Acquista, A. 2003. *The survival guide: What to do in a biological, chemical, or nuclear emergency.* New York: Random House.

The following technical article provides detailed analyses of various dirty bomb scenarios with recommendations for emergency responders in the early hours:

Harper, F. T., S. V. Musolino, and W. B. Wente. 2007. Realistic radiological dispersal device hazard boundaries and ramifications for early consequence management decisions. *Health Physics* 93 (1): 1–16.

The following report provides technical analyses for evaluating shelter and evacuation options after a hypothetical nuclear detonation incident in Los Angeles:

Brandt, L. D., and A. S. Yoshimura. 2009. Analysis of sheltering and evacuation strategies for an urban nuclear detonation scenario. SAND2009–3299. Sandia National Laboratories. Livermore, CA.

The following textbook contains proceedings of a 2004 Health Physics Society summer school and provides a wealth of information, especially for professionals:

Brodsky, A., R. H. Johnson, Jr., and R. E. Goans, eds. 2004. *Public protection from nuclear, chemical, and biological terrorism.* Madison, WI: Medical Physics Publishing.

10

PREPARING YOUR FAMILY EMERGENCY PLAN

OVERVIEW

In the United States, especially following the shocking terrorist attacks in 2001 and the devastating aftermath of Hurricane Katrina in 2005, the public has an increased sense of vulnerability. However, a serious gap remains between what are perceived as real threats and the degree of personal preparedness. As such, many people and their families are still not prepared to deal with an emergency—whether it is a man-made terrorism incident or a natural disaster. Lower income families are even more likely to be unprepared.* In fact, a majority of people rely on "just-in-time" preparedness. In other words, if they are given sufficient warning of an impending disaster, only then will they start making plans and take action to obtain or organize what they need to have in their homes or what they need to evacuate quickly and safely.

This approach is questionable—even in the case of disasters where advance warnings can be provided, such as with hurricanes or wildfires. For disasters with little or no advance warning—such as nuclear or radiological emergencies, earthquakes, or tornadoes—such a "just-in-time" plan of family preparedness is of little value.

Some people feel that it is impossible to prepare for unexpected events. In the case of a nuclear emergency, they may feel that any attempt at preparedness is futile anyway. The fact is that taking simple preparedness actions helps all of us deal with all types of emergencies much more effectively, including

* Since 2002, the National Center for Disaster Preparedness, Columbia University Mailman School of Public Health, has conducted annual surveys of the American public's attitudes, perceptions, and personal preparedness in the context of disaster response. These reports can be obtained from www.ncdp.mailman.columbia.edu/index.html and detailed survey data are available upon request.

a major nuclear disaster. What it takes to prepare as a family does not take much time, can be done gradually with any size budget, and is useful for any type of emergency.

A wealth of useful information on family emergency preparedness is already available for free, and a number of sources are referenced in the "Resources" section at the end of this chapter. It is not our intent merely to duplicate that material here. Rather, we will cover highlights and emphasize more critical components—especially for a radiation emergency.

THE FAMILY EMERGENCY PLAN

A family emergency plan is essentially about thinking ahead and deciding what actions you and your family will take if confronted with an unexpected emergency, including:

- communication issues if your family is separated;
- food, water, and medicine, if unavailable for a period of time;
- evacuation routes, what you would need to take with you, and where you would go;
- planning for members of the family who have special needs, such as the elderly, the disabled, infants, or pregnant women; and
- your pets' needs—whether you stay at home, evacuate, or move to a shelter.

You should start by thinking about the types of disasters that are most likely to happen where you live. In your area, it could be winter storms, floods, or tornadoes. If you are not sure, you can contact your local emergency management office or American Red Cross chapter to find out the disasters more likely to occur in your area and request information on how to prepare for them. Make your plans accordingly and gather the supplies you need to remain sheltered at home or to take with you in case of evacuation.

Your plan on how to respond to a radiation emergency will be an extension of your plan for natural disasters. If you live within a 10-mile radius of a nuclear power plant, you should be familiar with the sirens and other warning systems in place to alert you. The local emergency management agency can provide you with instructions specific to where you live (see the "Resources" section).

Communication and Getting Reunited

Your family may not be together when disaster strikes. Therefore, it is important to plan in advance and discuss how you will establish contact and get

Figure 10.1
A giant message board at the Reliant Center in Houston, Texas, for Hurricane Katrina evacuees to locate friends and loved ones, September 3, 2005. (Photo: Ed Edahl/FEMA.)

back together (Figure 10.1). If disaster strikes your area, it may be easier to make a long-distance phone call than to make a local call. Therefore, select an out-of-town family member or friend to serve as the central point of contact so that separated family members can establish contact through that person. Decide in advance who that family member or friend is and make sure that everyone in the family knows this and has the contact information.

The American Red Cross offers a "Safe and Well" Web site where people can enter their names and contact information after a disaster and leave messages for family members (https://disastersafe.redcross.org). But it is far more practical and immediate if you establish this contact directly through a predesignated family member or friend.

You should also decide in advance where you will reunite if you cannot establish communication in different situations. For example, you should decide on

- a local assembly point (e.g., your mailbox or end of a cul-de-sac) in case of a house fire;
- a location outside your neighborhood (e.g., local grocery store or landmark) in case you cannot return home; and

- a more distant location (e.g., Uncle Harry's house) if there is a major regional disaster and you and the family are separated.

Ideally, this information is written down with addresses and phone numbers and kept in a purse, wallet, or backpack for easy reference.

Keep in mind that, in a radiation emergency, it is important to stay informed and follow emergency communications from the authorities. It may be that your family's predesignated location to meet (e.g., Uncle Harry's house) is in the path of nuclear fallout and is not a suitable location. Follow emergency instructions and try to establish contact with family members at your earliest convenience.

Sheltering at Home

You may receive emergency instruction from authorities to keep shelter at home for a few days before the danger is over or before you can safely evacuate. Refer to Chapter 9 for more details on the best location inside a building for shelter and what precautions you need to follow in a radiation emergency. If you live in a high-rise building, you need to discuss with your building manager or superintendent what plans are in place to accommodate sheltering for all residents. In a single-family home, your basement offers the best protection in a radiation emergency. If you have no basement, a room at the center of the house may be an option. You may also consider an alternate nearby location to use as shelter. As part of your family plan, you should discuss with family members which room you should go to in case of an emergency. If possible, store your emergency supplies in that room (preferably the basement).

EMERGENCY KIT

Following any major incident, terrorism, or natural disasters, it is likely that basic services such as electricity, water, gas, or telephones may be unavailable for days or even longer. Transportation and supply lines (for anything from gasoline to groceries) will likely be disrupted as well. You may have to keep shelter in your house for a period of time, or you may have to evacuate at a moment's notice. In either case, there is no time for shopping or searching for supplies. It is best always to keep a supply of essential items in case you need to shelter at home or evacuate. This supply includes food, water, personal medication, and other items such as a battery-powered radio, flashlights, and batteries. A suggested list of these emergency supplies is provided in Appendix A.

Several points regarding the list in Appendix A are worth highlighting:

- One of the most important items to include in your emergency kit is a battery-powered radio with extra batteries. You can also keep a hand-crank radio, which does not require batteries. Weather alert radios have bands specifically for emergency alert communications. In radiation emergencies, it is of utmost importance that you stay tuned and receive up-to-the-minute emergency instructions.
- Generally, it is recommended that you keep a 3-day supply of food and water. Think of this as a minimum supply; a 2-week supply is preferable. If you have lost power, keep the freezer door closed. Use perishable items in your refrigerator first. Next, use what you have in the freezer. Then, use the canned and dry goods you have stored away. Do not use any food or water that was left in open containers because they may contain contamination.
- If your resources are limited, you do not have to assemble your emergency supplies overnight. Prepare a checklist and start with what you can, beginning with more essential supplies. Build up your supplies gradually until you complete your checklist.
- The best way to avoid waste is to establish a rotation schedule with perishable items in your emergency supply, such as canned foods, bottled water, medication, snacks, and batteries. Use these items from your emergency supply and replace them each time with a fresh supply.* As a rule of thumb, replace your stored food, water, and batteries every 6 months. Ask your doctor or pharmacist about storing a 2-week supply of prescription medications.

To-Go Kit

You should organize your emergency supplies so that you can quickly put them in your car in case of an evacuation. Your to-go kit does not have to be as extensive as what you need to shelter at home, but it should contain the essentials: a 3-day supply of food and water, a week's supply of medication, cash (because ATMs may not be working), copies of your important personal documents, a change of clothes for every member of the family, blankets, flashlights, and extra batteries. It pays to prepare this to-go kit in advance. You may keep an even smaller version (a mini-kit) in your vehicle at all times.

---

* The water in your hot-water tank and water pipes may also be used as an emergency water supply. See the joint publication of FEMA and the American Red Cross, Food, and Water in an Emergency in the "Resources" section.

There is no reason to go overboard with all of this planning. But as you start to assemble some of these emergency supplies, you and your family will feel more confident and secure, and you will be better prepared. See Appendix A for a suggested list of emergency supplies and the "Resources" section in this chapter for additional information.

Did you know? In the early 1960s, the list of recommended supplies for a home shelter was simple and similar to what is recommended today (e.g., canned food and water, flashlight, battery radio, and a first-aid kit). There was also a familiar presence of commercialism: Vendors offered consumers prefabricated shelters for about $1,500, some of which came with decorative picture windows painted on the walls. Some shelter manufacturers made false claims that their shelters were approved by the Office of Civil Defense. Among many ordinary survival kits on the market, there were a number of interesting items, such as a "lifesaving kit" containing a salve to cause radiation to bounce off the body harmlessly, "antiradiation pills," a "fallout suit" for $22, and a "burial suit" for $50, which was a polyvinyl plastic wrapper in case someone died in the shelter. The burial suit was advertised to contain chemicals to keep the odor down and it could also be used as a sleeping bag for the living.*

Do You Need a Radiation Detector?

If you search the Internet for nuclear survival tips, you may see advice regarding purchase of portable, handheld radiation detectors. You may also see Web sites ready to sell such instruments at relatively affordable prices. These instruments may be refurbished Civil Defense instruments from the 1950s or more modern radiation detection instruments. Operating most of these portable instruments is relatively easy. However, the choice of instrumentation, its operability, units of measure, and its sensitivity to the particular type of radiation present are among the factors that may lead an inexperienced user to an erroneous conclusion regarding radiation levels. Therefore, it is far more important to own a battery-powered radio or television to tune in emergency response instructions than to own a radiation detector.

If you choose to purchase a radiation detector, pancake Geiger–Muller detectors and ionization chambers are good instrument choices. You can consult with health physics experts in your state's radiation control program (see "Resources") to seek advice on a particular make, model, or set of instruments depending on your needs, concerns, and budget.

* *Time*. October 20, 1961. 78 (16): 21–26.

FAMILY MEMBERS WITH SPECIAL NEEDS

Family members with special needs and vulnerabilities include infants, pregnant women, homebound or disabled family members, and the elderly. These family members may require special supplies (ranging from diapers to oxygen tanks) or medication that is not already part of your basic emergency supply list. They may have limited ability for evacuation or movement in general. You need to consider these needs in your family's emergency preparedness plan. If you have a family member who lives in a nursing home, you should become familiar with emergency plans that the nursing home has in place and, if necessary, adjust your family emergency plans accordingly.

Pets

Pets are considered members of the family with special needs. You should include supplies for your pet (food, medication) in your emergency kit. See Appendix A for a pet emergency supply checklist and Chapter 9 for precautions regarding radioactive contamination of your pet. If you are evacuating, make sure that you take your pet and your pet's emergency supplies with you. As part of your family emergency plan, you should explore ahead of time what options you have for evacuation that can accommodate your pet. If you are going to a friend's or another family member's home, can you take your pet? There are pet-friendly hotels and you can keep a list of these facilities handy. Some hotels with "no-pets-allowed" policies may change them in case of emergencies.

Public shelters, such as those operated by the American Red Cross, do not accept pets and only allow service animals. If you have a service animal, keep a copy of your service animal's license in your emergency kit. You may need to provide that license to shelter staff if you evacuate to a public shelter.

It would be helpful to develop a buddy system with a trusted neighbor who would be willing to help evacuate your pet in case you are not home when evacuation orders are issued. Keep in mind that animals behave differently under stress. They may panic and even bite or scratch, hide, or try to escape. Keep dogs securely leashed, transport cats in carriers, and do not leave pets unattended where they may run off. Make sure that your pets are wearing collars with proper identification and contact information in case they get lost. Carry a recent photo of your pets (Figure 10.2).

Figure 10.2
Dogs displaced by Hurricane Ike are sheltered at the local center set up by the Humane Society, Galveston Island, Texas, September 17, 2008. (Photo: Jocelyn Augustino/FEMA.)

Resources

Several Web sites offer detailed information and useful tips for family emergency preparedness:

The Federal Emergency Management Agency (FEMA) offers an online emergency preparedness course for families called "Are you ready? An in-depth guide to citizen preparedness" (http://training.fema.gov/EMIWeb/IS/is22.asp). The content of this course can be downloaded or you can request a free hard copy from the FEMA publications warehouse (800-480-2520).

The U.S. Department of Homeland Security "Ready.gov" Web site is an information resource (www.ready.gov/america/index.html).

The American Red Cross has developed a Web site (www.prepare.org) with practical information on disaster preparedness, especially for vulnerable populations such as senior citizens, children, people with disabilities, and pet owners.

Food and water in an emergency (August 2004) is a joint publication of FEMA and the American Red Cross that includes valuable tips (www.redcross.org/images/pdfs/preparedness/A5055.pdf).

The American Veterinary Medical Association Web site hosts a page where you can find animal disaster plans and resources by state (www.avma.org/disaster/state_resources/default.asp).

You can find contact information for your local emergency management and homeland security departments at www.ready.gov/america/local/index.html

You can find contact information for your state radiation control program at www.crcpd.org

11

MEDICAL RESPONSE TO RADIATION EMERGENCIES

OVERVIEW

RADIATION INJURY IS RARE, ALTHOUGH CASES do occasionally occur in occupational settings. In the United States, the Radiation Emergency Assistance Center/Training Site (REAC/TS), headquartered in Oak Ridge, Tennessee, has been providing expert medical advice and consultation on the treatment of radiation injury for more than 30 years. In addition, REAC/TS provides a capability 24 hours a day, 7 days a week to respond anywhere in the world where medical assistance for treating radiation victims is needed.

Over the years, the REAC/TS team has accumulated a wealth of knowledge from responding to incidents around the world. Outside the United States, the World Health Organization (WHO) and the International Atomic Energy Agency (IAEA) have provided guidance regarding treatment of radiation injury. In countries with operating nuclear power plants, local hospitals are likely to have emergency plans in place for treatment of radiation victims. Moreover, these hospitals usually participate in periodically scheduled radiation emergency drills. Therefore, if a radiation incident (accidental or intentional) is limited in scope and only a small number of patients is affected, they will most likely receive prompt medical services.

However, if an emergency has an impact on or concerns a larger population, people are likely to go to hospitals in numbers that will quickly overwhelm the capacity of the medical community. Even worse, if a radiation emergency involves large numbers of victims with life-threatening injuries, many such patients may not even make it to hospitals in time or, due to lack of resources, patients may not receive the care they need once they are at the hospital. In such situations, difficult medical decisions need to be made to maximize the number of people who can receive care using available medical resources. In this sense, a large-scale radiation emergency is no different from a large-scale natural disaster, where the human toll may exceed medical capacity and similar disaster triage decisions will need to be made.

Not surprisingly, clinical staff, physicians, nurses, and emergency medical services may be apprehensive about the personal health risks posed by working with and caring for victims of radiation events. These concerns, expressed in various opinion surveys and focus group research, are normal and expected. Many health care personnel share these feelings of concern. Specialized training and participation in emergency training exercises will provide first responders and first receivers with the information needed to care for victims of a radiation emergency while protecting their own safety. Fortunately, protection against radiation and radioactivity is fairly straightforward. This important topic was described in Chapter 9.

In this chapter, a number of principles for treating radiation injury are discussed. Medical textbooks and the scientific literature contain considerable clinical guidance on treatment of radiation patients. Much of the pertinent clinical information is available free on the Internet from government agencies and medical organizations. It is not the intent of this chapter to reproduce that material. Instead, a number of these resources are listed at the end of the chapter for readers interested in more detailed clinical guidance. The main points of this chapter can be summarized as follows:

- In cases of combined injury, when a patient is exposed to radiation and is also suffering from trauma (burns, fractures, open wounds), it is more urgent to treat the trauma because radiation injury, if it develops, will take some time to present itself.
- A victim of radiation exposure does not present any risk to caregivers.
- A person contaminated with radioactive materials may cross-contaminate caregivers. However, the health risks from such cross-contaminations are small and can be further minimized or eliminated using standard precautions (mask, gown, gloves).
- To ensure the health and safety of health care providers, radiation protection (health physics) expertise is an indispensable component of an effective medical response to radiation emergencies at hospitals.
- If a patient is suffering from limb- or life-threatening injuries, medical stabilization is the top priority even in the presence of radioactive contamination.
- In mass casualty situations, when victims' needs exceed available medical resources, medical authorities should be prepared to make critical and difficult triage decisions to maximize the number of survivors.
- Health care providers need radiation awareness training and participation in training exercises to feel comfortable with their

medical responsibilities in a radiation emergency, especially in a mass casualty situation.

Chapter 12 describes a number of medications specifically used for treating radiation exposure or contamination. Chapter 13 describes how government organizations mobilize available resources to respond to a large-scale radiation disaster.

MEDICAL TREATMENT ON THE SCENE

Emergency medical responders are trained to adhere to protocols. On scene, they assess the situation, determine the extent of injuries, triage victims based on their medical and surgical needs, stabilize critically ill patients, and transport those in need to hospitals. In a radiation emergency, essentially the same standard procedures are used. Victims at the scene may be evaluated using the following criteria:

- Did they sustain any injuries, especially limb- or life-threatening injuries?
- Were they exposed to radiation?
- Are they contaminated with radioactive materials?
- Are they ambulatory?

The first criterion is the most important and immediate triage decision at the scene. In many radiation drills, while precautions are taken to control entry to the incident site tightly and time is taken to decontaminate every patient, critically injured mock patients die while waiting.

A fundamental principle for responding to critically injured radiation victims is that decontamination of such patients is not necessary prior to medical stabilization and provision of life-saving care. Standard precautions (mask, gloves) used as a barrier against blood-borne pathogens are similarly useful in protecting caregivers against cross-contamination with radioactive materials. Even if these protective measures are not used or are unavailable and some cross-contamination does occur, health risks to caregivers are minimal in nearly all cases.

There are some exceptions, however, when the work environment or the victims present an immediate health hazard to medical responders. First, the incident scene itself may be highly contaminated. Radiation exposure levels must be measured at the scene and any high-radiation sources (hot particles or fragments) identified and removed by qualified personnel. Second, some victims may have embedded radioactive shrapnel, exposing themselves and others to high radiation levels. Both of these situations can be easily and quickly evaluated using commonly available portable radiation detection instruments.

Injured patients who are contaminated but do not have life-threatening injuries can be cleaned before being transported or receiving medical care. Any open wounds must be properly irrigated with saline, and the wound needs to be shielded from fluid runoff from other parts of the body to reduce the chances of internal contamination.

In addition, it is likely that a large number of people at the scene may have been affected by the incident but do not need hospitalization. Some may need assistance with decontamination. Others may need medical assistance. But everyone will require information and instructions on what to do. This information and any instructions need to come from public health or emergency management authorities, and they need to be easy to understand and follow and available in languages appropriate for the community.

When the Affected Population at the Scene Is Small

What is meant by "small" in this context? The affected population is small when it does not exceed the capacity of local emergency response resources to provide on-scene assistance. Ideally, under such conditions, the following should take place:

Radiation monitoring should be offered at the scene to screen everyone for radioactive contamination.

All persons should be interviewed to determine possible radiation exposure.

Assistance with washing and decontamination should be provided before the affected population leaves the scene.

Contact information for follow-up medical assistance and any necessary sampling of blood, urine, or other tests should be recorded.

When the Affected Population at the Scene Is Large

In this situation, radiation screening, decontamination, and medical services cannot be offered at the scene for everyone. It is impractical to keep people waiting (in lines or otherwise) until such services can be offered. Health authorities may instruct people to go home after providing them with instructions to wash and clean and report later for radiation screening or medical monitoring. This is particularly prudent for people who have not suffered injuries.

Self-decontamination is straightforward, as explained in Chapter 9. Carefully removing external layers of clothing (coats, jackets, trousers) and placing the garments in a trash bag can eliminate most of the contamination. Providing people with plastic bags before they leave the scene can help facilitate this. Also, if no running water is available, using moist wipes (or baby wipes) to clean the hands and face can be very helpful in minimizing the chances of incidental ingestion or inhalation of radioactive material. If supplies are limited, children and pregnant women have priority.

TRANSPORTATION TO THE HOSPITAL

Patients with life-threatening injuries should be stabilized and transported to hospitals without delay, per protocol. If they are contaminated, patients can be wrapped in clean sheets before placement on stretchers or gurneys to minimize spread of contamination. Clothing can be removed for storage in a plastic bag prior to or during transport to the hospital. If cross-contamination of ambulances with low levels of radioactive contamination should occur, it is appropriate to continue using the same ambulances for response and patient transportation. Contaminated vehicles can be cleaned later, before they are returned to routine medical response.

People who do not need high-level medical care should be directed away from hospitals. Many communities have already planned to set up and staff alternate care sites where lower acuity patients can receive care.

MEDICAL TREATMENT AT HOSPITALS

The first 60 minutes or so after a traumatic injury are generally recognized by the medical community as a "golden hour" of opportunity. During this narrow time window, the lives of severely injured people may be saved if they are triaged to receive care rapidly. In a radiation emergency, the urgency to receive trauma care is even more acute. Victims of a radiation event may have what is called a "combined injury." This means that, in addition to exposure to a high dose of radiation, the patient suffers from a traumatic injury such as a burn, fracture, or open wound.

For patients with combined injury, trauma management takes on greater urgency than does treatment of radiation-related injuries and illnesses because radiation-related illnesses, if they occur, will not develop immediately. The patient's prognosis worsens significantly if traumatic injuries are not promptly treated. If any surgery is necessary, it should be performed within 36 and no later than 48 hours of radiation exposure. After this time, victims of radiation exposure will have increasingly lower levels of platelets and white blood cells, which will compromise wound healing and also make them more susceptible to infections.

It is important to keep the following in mind:

- Patients exposed to radiation—even those exposed to high doses and suffering from radiation sickness—do not pose any danger to others. The situation is analogous to a person who has received a CT scan. From the radiation point of view, he or she is not contaminated and not a danger to others. In an emergency, radiation doses to the person may be higher than that of a CT scan, but the concept is identical.
- Patients with external contamination on their clothes and bodies need to have that contamination removed. Complete decontamination is not always possible, however, because some radioactive materials can remain fixed to the skin surface or within a wound. Such residual contamination may pose a small risk to the individual, but does not pose a risk to caregivers.
- Patients with internal contamination (i.e., radioactive materials inside their bodies) pose no danger to others, although their sweat, urine, feces, and vomit will have some radioactive contamination and should be handled appropriately. This situation is similar to that of tens of thousands of patients who are treated every day with a variety of nuclear medicine procedures on an outpatient basis and sent home "contaminated." These patients' bodily fluids are also contaminated with radioactive materials, but pose no danger to family members as long as certain simple precautions are followed at home. Similarly, health care providers can follow simple steps to protect themselves (see Chapter 14). Generally, standard precautions (masks, gowns, gloves, and booties) will provide adequate protection. Radiation protection staff or health physicists at the hospital can ensure that health care providers and their work environments are properly protected. Radiation protection specialists at the hospital should work under supervision of the attending emergency services physician unless other reporting hierarchy is to be followed according to the hospital's radiation emergency response plan (Figure 11.1).

Did you know? In disasters, the first wave of patients arriving at the hospital does not generally include the sickest. The most mobile patients, often self-reporting, arrive first. Therefore, in any disaster, it is important for hospitals not to expend limited clinical resources and exhaust space and personnel by providing care to the first arrivals.

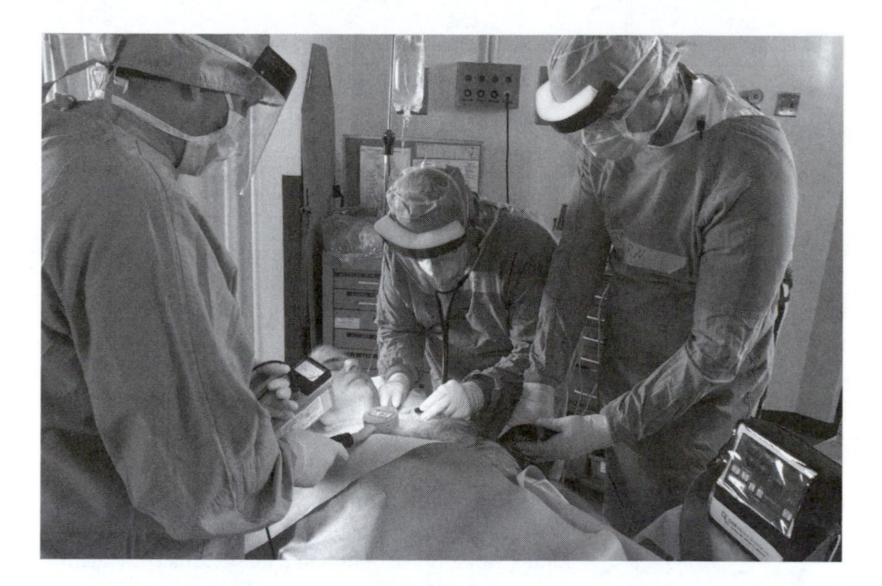

Figure 11.1
Clinicians and a health physicist (on the left) training together in a REAC/TS drill respond to a mock radiation incident. (Photo courtesy of REAC/TS Oak Ridge Associated Universities.)

Medical Treatment Options

Medical treatment for radiation injuries can range from a simple administration of antibiotics and intravenous (IV) fluids to complicated procedures such as bone marrow transplants (Table 11.1). The choice of treatment depends on the clinical presentation of a patient, available medical resources, total numbers of patients, and, ultimately, the medical judgment of the attending physicians. A number of available medical treatments are described here and additional resources for detailed clinical guidance are provided at the end of this chapter.

Supportive Care Supportive care includes routine clinical procedures such as administration of antibiotics (to fight infection),* antiemetic drugs (to help control vomiting), antidiarrheal drugs, analgesics (pain medication), and fluids and electrolytes. Although provision of this basic level of care does not require highly specialized medical facilities, it can save the lives of many victims of low to moderate doses of radiation.

* Antibiotics are given if there is evidence of infection; they are not generally given as prophylaxis.

Table 11.1 Treatment Options for Victims of Radiation Exposure

Treatment[a]	Description
Supportive care	Includes administration of fluids and electrolytes and drugs to treat or control infections, vomiting, diarrhea, or pain. Can be administered at alternate sites (away from hospitals). Can potentially save many lives.
Cytokine therapy	For patients with suppressed immune systems. Already has widespread clinical use to support patients undergoing chemotherapy or radiotherapy.
Decorporation therapy	For patients with *high* levels of internal contamination. These drugs are described in Chapter 12.
Transfusion	For patients with low blood counts. Blood supply may be short because large amounts of blood may be used initially to treat trauma patients.
Bone marrow transplant	For patients with nonresponsive bone marrow. Suitable for carefully selected patients with no other traumatic injury. Not a viable option in a mass casualty incident.
Palliative care	For patients with poor prognosis. Medication aimed at reducing pain and suffering.

[a] These treatments are for symptoms of radiation exposure only. Treatment of traumatic injuries (such as burns and wounds) takes priority.

Cytokine Therapy Cytokine therapy is aimed at helping the body recover its ability to fight infections. Cytokines are naturally occurring proteins whose function is to stimulate stem cells in the bone marrow to divide, thereby increasing production of needed blood cells. Each class of stem cells responds to a specific set of cytokines. Scientists have figured out how to manufacture such cytokines and deliver them as drugs to help patients with suppressed immune systems.*

One class of cytokines can stimulate production of neutrophils, the most abundant type of white blood cells. Cytokine therapy can help patients who are at risk for infection following a fall in the number of these infection-fighting white blood cells. Cytokine therapy is used routinely to support cancer patients who have low neutrophil counts as a side effect of radiation therapy or chemotherapy. Cytokine therapy has also been used to treat radiation sickness in a number of radiation accidents worldwide. See Chapter 12 for a more detailed discussion of this treatment method.

* In medical jargon, when the ability of the body's immune system to fight infections is reduced, the patient is said to be immunosuppressed. An immune system thus compromised presents an opportunity for infections. Such infections are referred to as opportunistic infections.

Transfusion Transfusion of blood cells and platelets is used for patients with severe bone marrow damage. The effect of radiation damage on the bone marrow will take some time, perhaps several days, to become critical. In the initial hours after a mass casualty radiation incident, transfusions will be needed to treat victims of traumatic, rather than radiation-induced, injuries. This situation can exhaust blood supplies rapidly. Therefore, every effort should be made to mobilize potential blood donors and replenish blood supplies for treatment of radiation-exposed patients with severe bone marrow damage, who will need the transfusions later.

Stem Cell (Bone Marrow) Transplantation Bone marrow transplant procedures usually support patients who receive high-dose chemotherapy. In addition to destroying cancer cells, high-dose chemotherapy destroys normal blood-producing stem cells in patients' bone marrow—similar to the damage caused by high-dose, whole-body radiation exposures. Bone marrow transplantation from a matching donor can restore the body's ability to produce normal blood cells. For victims of serious radiation exposure emergencies, there are a few caveats.

Where pockets of healthy blood stem cells remain, these cells can usually migrate and repopulate the bone marrow without the need for transplantation. Conversely, if the radiation dose to the whole body is high enough that it damages all of a person's bone marrow, chances are that other organs have also sustained critical damage. For these individuals, a bone marrow transplant may only serve to prolong life for a few weeks or months; it will not cure that individual. Given the poor quality of life that results from severe radiation exposure, using bone marrow transplants to delay the inevitable is an ethical decision that the physician, the patient, and the patient's family should make together. For carefully selected casualties who have no complicating injuries such as burns or trauma, bone marrow transplant should be a considered option. But, for the most part, medical response planners should not regard stem cell transplantation as a procedure that can save large numbers of lives in a mass casualty radiation emergency.

Palliative Care Palliative care is aimed at relieving symptoms and reducing suffering, but not necessarily curing a health condition. Victims of radiation events who develop multiple organ failure within hours of exposure to whole-body doses exceeding 10 Gy (1,000 rad) have no chance for survival. In triage terms, they are in the "expectant" category: They have a poor prognosis and are not expected to survive, even with treatment. Rather than aggressive treatment, appropriate and ethical management of these patients includes the use of comfort measures such as drugs to alleviate pain or manage vomiting and diarrhea. Attention should be paid to counseling needs and spiritual care for the patients as well as their families and friends.

Laboratory Tests

People who work with radiation as part of their job wear a device that measures or records their daily external radiation dose. If there is an accident, the same device (called a personal dosimeter) records the dose received from the accident. This technology has been around for decades and is used routinely by workers in the nuclear industry and by the radiology staff in all hospitals. People who work with radioactive materials are also routinely tested to establish their baseline levels of natural radioactivity and to measure the amount (if any) of radioactive contamination that may have entered their bodies.

In a radiation emergency, members of the general public may be exposed to varying doses of external radiation and internal contamination may be a possibility. Information about the doses that may have been received or how much contamination may be inside bodies will be helpful information for public health and medical authorities in managing the affected population.

In a limited-scale radiation incident, medical authorities will seek to collect as much information as possible to guide the best medical management for their patients. Clinical data (e.g., blood and urine test results), information about details of the incident (the location of the individual relative to the event), and other relevant health-related information can be used in aggregate to decide on the best course of treatment or whether treatment is even necessary.

Following a mass casualty incident, it is unlikely that such detailed information will be available for each person. Clinicians will need to use their best professional judgment to treat patients based on presenting signs and symptoms. For example, the time to emesis (vomiting) is a potentially useful indicator of radiation sickness (Table 6.2 in Chapter 6). Did the patient vomit? How long after exposure did the vomiting start? Persons who do not experience vomiting in the first 2–4 hours after exposure are more likely to have received a sublethal dose of radiation. That is, assuming no additional injuries, these individuals should survive.

Conversely, onset of nausea and vomiting within an hour or two is indicative of an exposure to high and dangerous doses of radiation. Of course, stress and anxiety alone can cause nausea and vomiting, so although these clinical markers are useful, they are not 100% reliable.

Well-established laboratory tests can be used to determine whether people have been exposed to radiation or are internally contaminated with radioactive materials (Table 11.2). Three tests can be used to evaluate or assess exposure to radiation: (1) complete blood count with differential, (2) cytogenic biodosimetry, and (3) electron paramagnetic resonance. These tests are

Table 11.2 Laboratory Tests to Estimate Radiation Dose or Extent of Internal Contamination

Laboratory Test	Consideration
For Estimating Radiation Dose	
CBC with differential	Commonly available at most clinical laboratories. Quick and inexpensive. Not sensitive at low doses, but useful for guiding early medical decisions; useful in mass casualty incidents.
Cytogenetic biodosimetry	The most accurate and sensitive method for determining radiation dose. Considered the "gold standard." Only a limited number of laboratories can do the analysis. Takes a minimum of 3 days to process samples.
Electron paramagnetic resonance (EPR)	Currently, not as sensitive as cytogenetic biodosimetry. Potentially powerful tool for retrospective biodosimetry, but currently has no clinical application. Use in radiation emergencies under development.
Field kits	Under development. Potential use for field triage purposes. Uses small volume of blood to assay for gene expression in response to radiation.
For Estimating the Extent of Internal Contamination	
Urine or fecal bioassay	Well-established methodology. Early results can help determine if decorporation therapy is needed. The only reliable test of internal contamination with alpha emitters.
Whole-body counting	No need for sample collection. Used for internal contamination with gamma emitters. Offers less uncertainty than urine or fecal assays. Fixed facilities and mobile units are available.
Nasal swabs	Useful only if radioactive material is inhaled. No reliable quantitative information. Limited time window to collect sample. Useful in occupational settings, but applicability to mass casualty incidents is highly questionable.

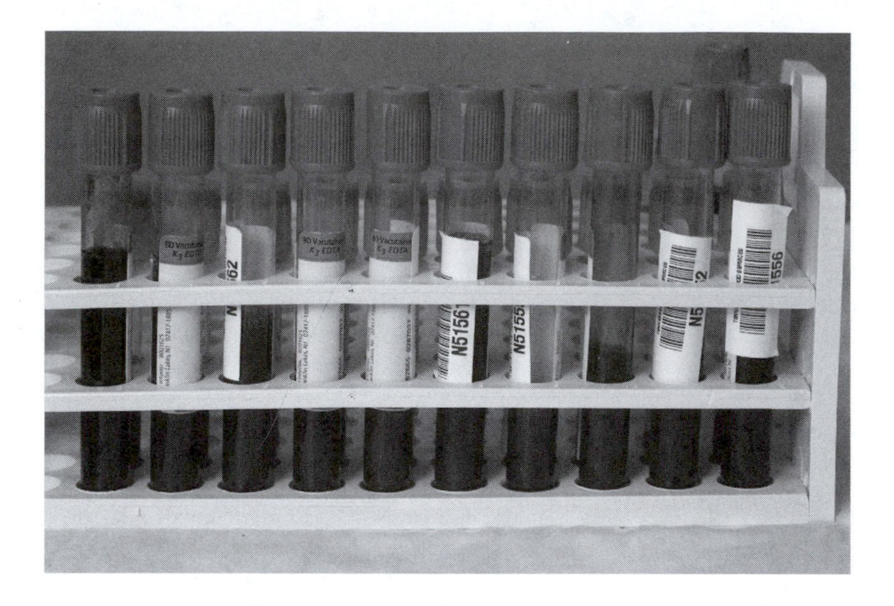

Figure 11.2

Laboratory analyses of blood samples can help clinicians assess the extent of radiation exposure for a patient. The rate at which the lymphocyte (a type of white blood cell) count falls and how low it gets correlates well with the radiation dose received.

briefly described in the following sections. The following tests can be used to assess the amount of internal contamination: (1) urine or fecal bioassay, (2) whole-body counting, and (3) nasal swabs. With the exception of the complete blood count with differential, completing these tests requires specialized equipment and laboratories.

Complete Blood Count (CBC) with Differential CBC with differential is a commonly available blood test that can be run quickly and inexpensively at almost any clinical laboratory (Figure 11.2). For a suspected radiation victim, it is important to obtain a baseline CBC as quickly as possible and then collect subsequent samples every 4–6 hours. In particular, health care providers will want to pay careful attention to the circulating lymphocyte (a type of white blood cell) count.

The rate at which the lymphocyte count falls (and how low it gets) correlates well with the radiation dose received. This information is useful to clinicians deciding on a course of treatment. Other laboratory tests described in this section require specialized equipment and trained staff as well as a longer time to obtain results. Therefore, if the number of victims is large, the CBC with differential may be the only laboratory test immediately available to clinicians to assess the extent of radiation exposure, a given patient's prognosis, and a possible course of treatment.

Cytogenetic Biodosimetry "Cyto" is the Greek word for cell and "genetics" refers to the DNA molecules and chromosomes within these cells. Biodosimetry means measuring the dose to a living organism. Cytogenetic biodosimetry is the most accurate and sensitive method for determining the amount of radiation dose absorbed by a patient's body. This method has been around for decades and is considered the "gold standard."

In cytogenetic biodosimetry, a blood sample is collected and lymphocytes within that sample are isolated and grown in culture. After some processing, cells are examined under a microscope and the number of chromosome damages in that group of cells is counted. The higher the dose is, the greater are the number of chromosome damages (Figure 11.3). The optimal time for collecting blood samples for this analysis is 24 hours after exposure to radiation.

This is a sophisticated analysis and cannot be done at a hospital. Only a limited number of laboratories around the world can do this analysis with acceptable standards. Furthermore, because samples take at least 3 days of processing, results are not immediately available. When available, results of cytogenetic analyses are invaluable in guiding patient management. In the United States, the Cytogenetics Biodosimetry Laboratory (CBL), which is part of the Radiation Emergency Assistance Center/Training Site in Oak Ridge, Tennessee, was reestablished in 2007 with improved capacity to analyze such samples in a large radiation incident (Figure 11.4). The Armed Forces Radiobiology Research Institute (AFRRI) in Bethesda, Maryland, and similar laboratories in Germany, Canada, and Japan collaborate with CBL in Oak Ridge to standardize the methodology further.

Electron Paramagnetic Resonance Also referred to as electron spin resonance (ESR), electron paramagnetic resonance (EPR) is a sophisticated technique that measures unpaired electrons or free radicals produced by ionizing radiation. Using this technique, it is possible to evaluate an individual's bone or teeth (noninvasively), nail, hair, or even articles of clothing worn at the time of exposure to estimate radiation doses. The EPR technique is currently not as sensitive as cytogenetic biodosimetry; however, for triage purposes in the initial medical response, high sensitivity is not needed. Currently, EPR does not have a clinical application but will perhaps in the future. The technique can be used to reconstruct radiation exposure incidents or as a forensic tool.

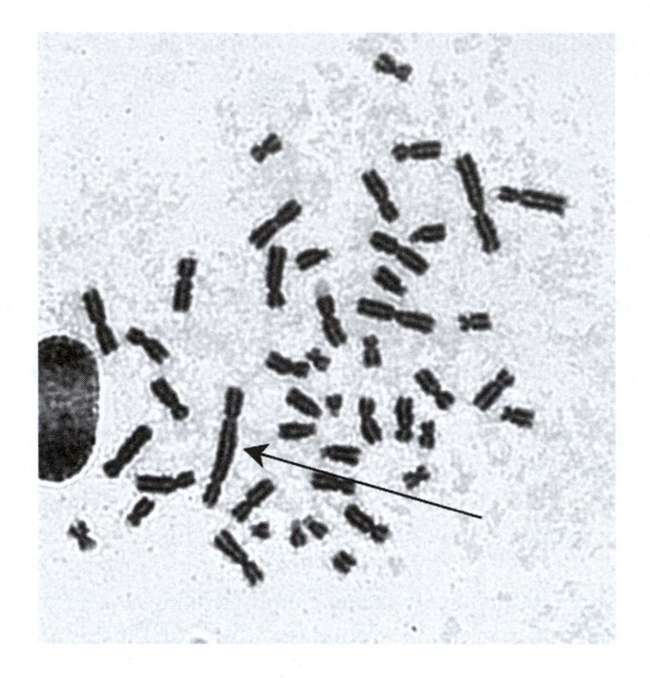

Figure 11.3
Cytogenetic biodosimetry is currently the most sensitive and accurate method of determining the radiation dose a person has received. Chromosome aberrations that are scored in this analysis are dicentric chromosomes. A dicentric chromosome (shown with arrow) has two centromeres. (Photo courtesy of REAC/TS Oak Ridge Associated Universities.)

Urine or Fecal Bioassay For these tests, a sample of the patient's urine or feces is analyzed in a specialized laboratory to determine the amount of radioactive contaminant in the sample (Figure 11.5). This information will then be used to estimate the amount of radioactive contamination in the body. This is valuable information when internal contamination is suspected. The scientists (internal dosimetrists, to be exact) have to make certain assumptions about the patient's metabolism to calculate the amount of contamination in the body. Therefore, some uncertainty is always associated with the estimated amounts of internal contamination, especially in samples collected shortly after an incident occurs. This uncertainty is reduced with subsequent follow-up samples.

Early results, however, do provide valuable information that physicians can use to determine whether decorporation therapy is needed (see Chapter 12). When the contaminant is a gamma-emitting isotope, there are alternatives

Figure 11.4
Dr. Gordon Livingston of REAC/TS is analyzing chromosome images next to an automated workstation that can locate the cells and scan images from the slides continuously. The number of dicentric chromosomes is compared to calibration curves to estimate a radiation dose. (Photo courtesy of REAC/TS Oak Ridge Associated Universities.)

to urine or fecal bioassays; the level of gamma radiation emanating from the patient's body can be measured and used to estimate the amount of contaminant in the body (direct bioassay). When the contaminant is an alpha emitter (Chapter 3), a urine or fecal bioassay is the only reliable test to determine internal contamination.

Whole-Body Counting In whole-body counting—another method of determining the amount of internal contamination—no samples are collected from the individual. A machine detects and analyzes the radiation emanating from the body and the analyst can see how much of that radiation is from natural sources and how much can possibly be from contamination. When the contaminant is a gamma emitter (Chapter 3), whole-body counting offers an advantage over a urine or fecal bioassay because radioactive contaminants in the body are measured directly and fewer assumptions are made to estimate the amount originally inhaled or ingested by the patient. A number of facilities offer whole-body counting and health authorities can also deploy mobile units.

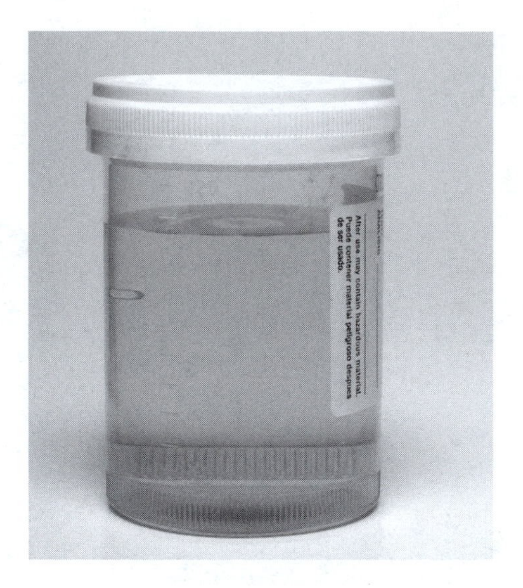

Figure 11.5

When a person is internally contaminated with radioactive materials, a fraction of the contamination is excreted through urine. Laboratory analysis of urine samples can detect minute amounts of radioactive contamination and that information is helpful in estimating the total amount of radioactive contamination in a person's body.

There are also methods by which gamma cameras used routinely in most hospitals for imaging can be converted and used as whole-body counters in a radiation emergency. This procedure requires that a skilled medical physicist be on the hospital staff. When the contaminant is an alpha emitter, alpha particles emitted inside the body cannot be measured outside the body; thus, whole-body counting is not useful. For internal contamination with alpha emitters, then, a urine or fecal bioassay is the only reliable test.

Nasal Swabs This test involves collecting a sample from each nostril onto a cotton swab to analyze for radioactivity. In occupational settings, when it is suspected that a worker may have inhaled radioactive materials, nasal swab samples can offer useful information. But in a mass casualty radiation emergency, the usefulness of such a test is questionable. Nasal swabs can be useful during a limited time window after inhalation. Negative results do not confirm that a person did not inhale radioactivity. A positive sample may indicate inhalation, but it offers no reliable quantitative information. If radioactivity was ingested rather than inhaled, a nasal swab would offer no information. Furthermore, collecting nasal swab samples from each person and handling and processing these samples take valuable staff resources. In a mass casualty situation, scarce resources should not be used for collecting nasal swab samples that have limited use.

Current Research and What to Expect Currently, no field-deployable rapid test can be performed to assess the radiation dose received and help in triage and management of patients. Such test kits, however, are feasible. Exposure to ionizing radiation produces dose-dependent changes in the expression of many genes, potentially providing a means to assess radiation exposure and to quantify dose. Sensitive test kits can use a small volume of blood to make such dose assessments (based on gene expression) in a short period of time. The U.S. government is supporting research in this area, and field-deployable, rapid clinical test kits may be developed and become available in the near future. Clinical applications of EPR (or ESR) techniques may also be developed; this will be similarly useful in quick assessment of radiation dose for each affected individual before any clinical symptoms appear.

MANAGING RESOURCES

Hospitals with effective emergency response plans should make provisions for what is referred to as "graceful degradation" of care. This means that, in a mass casualty situation when resources are stretched beyond their limits, hospitals need to alter how they provide care. Hospitals need to adapt and continue to provide care to increasing numbers of patients. For example, patients may need to be treated in hallways, where their privacy is compromised but their wounds are treated.* The goal of graceful degradation is to avoid a catastrophic collapse of health care delivery. See "Resources" for more information about altering standards of care in mass casualty situations.

Another resource management issue relates to training. Many radiation training courses with continuing education credits are available for health care professionals. However, some clinicians may opt not to take these training courses because they consider the chances of facing a radiation emergency to be low. Understandably, they choose to invest time taking continuing education courses that address more immediate needs in their medi-

* The Joint Commission on Accreditation of Healthcare Organizations (JCAHO). 2003. Health care at the crossroads: Strategies for creating and sustaining community-wide emergency preparedness systems (available from www.jointcommission.org).

cal practices. Hospitals can address this apparent shortage of clinical subject matter experts in two ways:

- Hospitals can take better advantage of radiation expertise already in their facilities. These include the staff working in the radiation safety office, nuclear medicine, radiology, radiation oncology, and hematology departments. These professionals should be consulted and included in the hospital's radiation emergency response plan.
- Hospitals can establish a network of outside radiation experts to serve as consultants in a radiation emergency.

Health Care at Alternate Locations

As mentioned earlier, it is important to keep people who do not need hospitalization away from hospitals. In their emergency response plans, many communities have incorporated provisions to supply limited health and medical care at alternate locations. In various jurisdictions, these facilities may be called alternate care centers, acute care centers, neighborhood emergency help enters, medical care points, points of distribution, or community reception centers. These alternate locations may be set up in area high schools, National Guard armories, airport hangers, community centers, or even in houses of worship. They may be used to provide radiation screening, counseling, assistance with washing and decontamination, medication, medical care, transportation to hospitals if needed, or information to the public. The critical importance of establishing these sites is twofold: (1) to serve the public in a timely manner, and (2) to keep the public from going to hospitals when there is no need.

DISASTER TRIAGE

Triage is the process of assessing and prioritizing patient needs and evaluating available resources to meet those needs; the ultimate goal is to maximize the number of survivors. In a disaster, triage is a dynamic process as the number of patients and availability of resources change with time. With sufficient clinical staff, adequate facilities, ample medical supplies, and enough transportation services to meet patient needs, the process of triage is rather simple. In a disaster situation, however, effective triage practices are based on doing the greatest good for the greatest number of people. This involves making difficult decisions: People who have a better chance of survival should have a higher priority at the expense of those with a lower chance of survival. The principles of disaster triage apply to large-scale radiation emergencies as well.

Because the process of triage does not involve a full medical evaluation, health care providers must rely on a person's presenting signs and symptoms. The START (simple triage and rapid treatment) system places adult trauma victims into one of four categories on the basis of rapidly obtained physical findings, such as pulse and respiratory rate and overall level of consciousness. This triage system has four categories:

1. *Immediate* refers to patients who can be saved if *immediately* transported to receive available medical care.
2. *Delayed* refers to patients who have suffered injuries, even significant injuries, but whose transportation can be delayed for up to several hours without significantly affecting their condition.
3. *Minor* refers to patients with only minor injuries. These are sometimes referred to as the "walking wounded" and may be able to administer self-help.
4. *Expectant* refers to patients who are unlikely to survive because of severity of their injury or lack of available care. In a mass casualty situation, victims of high-dose radiation exposure who also suffer from severe traumatic injuries such as burns and wounds are likely to be placed in the expectant category. In disaster triage, this group of patients has the lowest priority.

Resources

The radiation event medical management (REMM) Web site is an excellent online resource with medical guidance for physicians about clinical diagnosis and treatment during mass casualty radiation emergencies. Much of the information is also helpful for nonclinicians. This site was created by the U.S. Department of Health and Human Services, Office of the Assistant Secretary for Preparedness and Response, in cooperation with the National Library of Medicine (www.remm.nlm.gov).

The U.S. Centers for Disease Control and Prevention has produced a number of useful tools for clinicians and hospitals that can be downloaded from the CDC Web site or ordered free of charge. These include a pocket guide for clinicians; a 16-minute, just-in-time training video for managing radiation patients; more in-depth discussion and training videos; hospital guidelines; and fact sheets. Many of these products are accredited for continuing education and can be downloaded (http://emergency.cdc.gov/radiation/clinicians.asp).

In the United States, REAC/TS provides week-long training courses for clinicians on how to treat radiation injury. These courses are accredited for continuing education credits and include hands-on exercises. The courses are offered for a nominal fee in Oak Ridge, Tennessee, but organizations can request on-site training as well. Physicians, physician assistants, nurses, emergency medical technicians, health physicists, and first responders can all benefit from these courses (http://orise.orau.gov/reacts/courses.htm). REAC/TS also has a number of valuable resources on medical response to radiation on its Web site (http://orise.orau.gov/reacts/).

Regarding medical guidelines for treating radiation injury, several publications are available. The following are a number of useful and relatively brief publications:

IAEA. 1998. *Planning the medical response to radiological accidents.* Vienna: IAEA (accessible from http://www-pub.iaea.org/MTCD/publications/PDF/Pub1055_web.pdf).

The U.S. Department of Homeland Security Working Group on Radiological Dispersal Device (RDD) Preparedness May 2003 report (accessible from http://www1.va.gov/emshg/docs/Radiologic_Medical_Countermeasures_051403.pdf).

Waselenko, J. K., T. J. MacVittie, W. F. Blakely, N. Pesik, A. L. Wiley, W. E. Dickerson, H. Tsu, et al. 2004. Medical management of the acute radiation syndrome: Recommendations of the strategic national stockpile radiation working group. *Annals of Internal Medicine* 140 (12): 1037–1051.

Flynn, D. F., and R. E. Goans. 2006. Nuclear terrorism: Triage and medical management of radiation and combined-injury casualties. *Surgical Clinics of North America* 86:601–636.

For a more in-depth discussion of when stem cell transplantation may be used in treating radiation injury, see:

Densow, D., H. Kindler, and A. E. Baranov. 1997. Criteria for the selection of radiation accident victims for stem cell transplantation. *Stem Cells* 15 (suppl 2): 287–297.

Weidorf, D., N. Chao, J. K. Waselenko, N. Dainiac, J. O. Armitage, I. McNiece, and D. Confer. 2006. Acute radiation injury: Contingency planning for triage, supportive care, and transplantation. *Biology of Blood and Marrow Transplantation* 12 (6): 672–682.

Hematologists and oncologists may be particularly interested in two publications describing their role in patient management and the

Radiation Injury Treatment Network (RITN), which comprises transplant centers, donor centers, and cord blood banks spread throughout the United States (http://bloodcell.transplant.hrsa.gov/ABOUT/RITN/index.html):

Weinstock, D. M., C. Case, J. L. Bader, N. J. Chao, C. N. Coleman, R. J. Hatchett, D. J. Weisdorf, and D. L. Confer. 2008. Radiologic and nuclear events: Contingency planning for hematologists/oncologists. *Blood* 111 (12): 5440–5445.

Dainiak, N., J. K. Waselenko, J. O. Armitag, T. J. MacVittie, and A. M. Farese. 2003. The hematologist and radiation casualties. *Hematology American Society of Hematology Education Program* 2003:473–496.

The following two textbooks on medical management of radiation accidents have been written by several authors describing a number of radiation accident case studies:

Gusev, I. A., A. K. Guskova, and F. A. Mettler, eds. 2001. *Medical management of radiation accidents,* 2nd ed. Boca Raton, FL: CRC Press.

Ricks, R. C., M. E. Berger, and F. M. O'Hara, Jr., eds. 2002. *The medical basis for radiation-accident preparedness.* Boca Raton, FL: Parthenon Publishing Group.

The following report has an informative section (Section 4) devoted to medical management of radiation casualties: National Council on Radiation Protection and Measurements (NCRP). 2001. *Management of terrorist events involving radioactive material.* NCRP report no. 138. Bethesda, MD.

The Armed Forces Radiobiology Research Institute provides radiation exposure assessment software called BAT (biodosimetry assessment tool). Clinical information from patients can be input and stored. The program uses all available clinical information to provide a best estimate for the patient's dose. Software can be requested, free of charge (http://www.afrri.usuhs.mil/outreach/biodostools.htm).

The REMM Web site also provides an online dose assessment tool. This resource is very simple to use. The user (physician) can input the time to onset of vomiting or results of the patient's blood or cytogenetic tests and, by the click of a button, obtain an estimate for the radiation dose and suggested treatment protocol. However, this utility cannot store patient information (http://remm.nlm.gov/ars_wbd.htm).

12

RADIATION DRUGS

OVERVIEW

THE TOPIC OF RADIATION DRUGS INTERESTS many cross sections of the community, including first responders, clinicians, and public health professionals, as well as members of the general public. Yet, there are often misconceptions about these drugs. Do they provide immunity against radiation? Do they cure the effects of radiation injury or provide an antidote for it? How available are they and how do people obtain them? What if people cannot find them? Just a sampling of news headlines in recent years may explain why the general public and professionals alike might have such questions and how misconceptions can arise from sensational coverage of this topic:

- "Approval of radiation drug sought by Pentagon" (*New York Times,* December 7, 2001)
- "Feds stockpiling antiradiation pills; drug called 'psychological valium'" (*Atlanta Journal Constitution,* January 13, 2002)
- "FDA wants antiradiation drug" (*Wired,* February 1, 2003)
- "Paint pigment Prussian blue may be radiation antidote" (Associated Press, January 31, 2003)
- "Antiradiation pills are urged for children" (Associated Press, April 7, 2003)
- "Scientists have created a drug which they say could protect people from the effects of a nuclear attack" (BBC News, April 8, 2003)
- "Radiation sickness drug developed" (*Washington Post,* May 19, 2003)
- "Radiation sickness drug could save thousands" (*New Scientist,* May 20, 2003)
- "FDA approves dirty-bomb antidote" (Associated Press, October 17, 2003)
- "U.S. drops plans for antiradiation drug" (*New Scientist,* March 17, 2007)
- "Drug to protect against radiation" (BBC News, April 10, 2008)

In fact, health care providers may use several classes of drugs to treat individuals exposed to high doses of radiation or who have large amounts of radioactive contamination in their bodies. These drugs belong to one of three general categories:

1. palliative care or supportive care drugs used to relieve pain, complications, and side effects of radiation sickness;
2. drugs used to treat exposure to radiation specifically; and
3. drugs used to block the absorption of radioactive contaminants into body organs or help excrete or eliminate them from the body.

In the previous chapter, we described general palliative and supportive care for victims of radiation injury. This care includes, but is not limited to, antiemetics (to help control vomiting), antidiarrheals, intravenous fluid replacement, blood product transfusions, and antibiotics. In this chapter, we concentrate exclusively on categories of drugs used specifically to treat radiation exposure and contamination, the so-called "radiation drugs" (Table 12.1). U.S. Food and Drug Administration (FDA)-approved radiation drugs will receive added emphasis. We will also mention a number of other drugs that may be in various stages of development, and describe their effectiveness and limitations without overusing clinical jargon. Resources are provided for additional information.

It is important to recognize that there is no radiation panacea. No currently available radiation drug offers immunity against radiation exposure or magically removes radioactive contamination from the body. The information provided in this chapter should help explain when and how such drugs may be useful.

Not every radiation emergency will require distribution of radiation drugs nor will all people involved in a radiation emergency necessarily need treatment with such drugs. The single most effective strategy in any radiation emergency is for the public to take protective actions to help avoid or minimize exposure and contamination in the first place. Why? Because no drug is as effective at reversing the effects of exposure or blocking uptake of contaminants as preventing their occurrence in the first place is.

DRUGS TO TREAT RADIATION EXPOSURE

Two distinct groups of drugs may be of use in case of exposure to high doses of radiation. One group, the radio-protective drugs, or radioprotectants, acts by preventing or limiting radiation damage to cells. The second group of drugs may help damaged tissues recover *after* radiation injury has occurred.

Table 12.1 Common Radiation Drugs

Type	Available for Emergency Use[a]	Description
For Irradiation (External Exposure)		
Radioprotectants	None	These drugs would offer broad protection against radiation injury. Currently, no such drug is suitable for emergency use, but the search continues.
Colony stimulating factors (CSFs)	Filgrastim	These drugs stimulate bone marrow stem cells to grow and differentiate into a variety of mature blood cell types and are useful in recovery from radiation injury to bone marrow. Filgrastim stimulates production of neutrophils (white blood cells), reducing chance of infection in radiation victims. It is available under the brand name Neupogen®. Administered by intravenous (IV) or subcutaneous (SC; under the skin) injections.
	Pegfilgrastim	The long-acting form of Neupogen. Available under the brand name Neulasta®.
For Radioactive Contamination		
Blocking agent	Potassium iodide (KI) or potassium iodate (KIO_3)	This drug protects the thyroid gland from uptake of radioactive iodine. It is available in tablet form and flavored liquid. Brand names include Iosat®, ThyroSafe®, and ThyroShield® (liquid form).
Decorporating agents	Prussian blue	These types of drugs help in excretion of radioactive materials from the body. Prussian blue helps to eliminate radioactive cesium or thallium. It is available under brand name Radiogardase™. It is a capsule.
	DTPA	This helps in elimination of radioactive americium, plutonium, or curium. It comes in two chemical forms, a calcium salt (Ca-DTPA) and a zinc salt (Zn-DTPA). Intravenous injection is the recommended method of administration.

[a] Refer to the text describing each type of drug to see which drugs are FDA approved and which could be made available under an emergency-use authorization.

Radioprotectants

Scientists have been searching for decades to find drugs that can protect body tissues from radiation damage. The motivation for this search has come from two arenas: military (for fight in nuclear wars) and health care (for the fight against cancer). During the cold war, soldiers faced the possibility of operations in a nuclear battlefield. Any drug providing protection against radiation damage would increase the troops' combat effectiveness. In the United States, the Armed Forces Radiobiology Research Institute (AFRRI) has been a leader in such research.

Similar research interests exist in the field of cancer therapy. When patients undergo chemotherapy or radiation therapy, the objective is to maximize the killing of tumor cells. But destroying tumor cells is always at the expense of and limited by damage to the patient's healthy tissue surrounding that tumor. Any drug that can protect the normal healthy tissue (or sensitize tumor cells) during treatment would offer what is called a "therapeutic advantage" in treating cancer.

> *Terminology:* Because these types of drugs offer their protective effects at the cellular level (i.e., they protect the inside of the cells from the harmful effects of radiation), they are also called cytoprotective agents.

During decades of research, scientists have identified cytoprotective agents that can be taken up inside cells and protect them against free radical damage produced by radiation (see Chapter 6). However, the use of such drugs is limited by their side effects. At doses of drug necessary to protect against acute radiation injury, side effects such as nausea, vomiting, and hypotension (low blood pressure) limit their use. Furthermore, for these drugs to offer any protective value, they have to be taken *before* exposure to radiation. This fact has two implications. First, the drugs are not useful for the purpose originally intended (i.e., on a nuclear battlefield). The combat effectiveness of vomiting, nauseated, and hypotensive troops is highly questionable and the entire approach would be unethical. Second, in the context of radiation emergencies, because terrorism incidents occur without notice, administering such drugs to a large population as prophylaxis is not possible.

These cytoprotective drugs, however, may have applications in radiation therapy treatments. One such drug that has been studied extensively is amifostine. It is used in clinical practice under the trade name Ethyol®. There is some evidence that amifostine helps prevent treatment-related secondary cancers, which is a particular concern for children in pediatric oncology. Treatment-related secondary cancers occur when a patient (usually a child or young adult) develops a new and different cancer, many years after the initial cancer, as a consequence of the therapy given to treat the initial cancer. For patients with life-threatening cancers in whom radiation exposures are planned in advance, the use of cytoprotective drugs may provide a therapeutic advantage, and the side effects of nausea and vomiting can be anticipated, planned for, and treated.

> *Did you know?* Before being commercially available, amifostine or Ethyol® was known simply as WR-2721 in scientific literature. "WR" stands for Walter Reed because this drug was originally a product of classified military research. Never found to have any military application, it is now approved by the FDA for use as a protectant of normal tissues during radiotherapy of head and neck cancers.

In summary, an effective radioprotective drug that can be used *after* a radiation incident to protect broadly against radiation damage is still elusive. Furthermore, in cases of acute exposures in a radiation emergency, because injury occurs at the cellular level within a very short period of time (less than a second), the concept of broad cellular protection using prophylactic drugs is not medically feasible. Drugs that can help the body recover faster, after radiation injury has already been inflicted, offer a more practical—although less desirable—approach for managing radiation injury.

Treating Radiation Injury

Rapidly dividing cells are the most susceptible to radiation injury. One organ containing rapidly dividing cells is the bone marrow—specifically, the stem cells that give rise to mature blood cells. This is the reason why clinical findings of radiation sickness include sharp drops in neutrophils and platelets (see Chapter 6). Persons with significantly low levels of neutrophils (<1,000/mm^3) have a depressed or compromised ability to fight infections and can die in the absence of proper treatment. A significantly low level of platelets (<10,000/mm^3) impairs the clotting process and severe life-threatening internal bleeding can result.

An effective medical approach to this problem has been used routinely as adjunctive therapy in the treatment of cancer patients undergoing aggressive chemotherapy. Chemotherapy, like radiation, affects rapidly dividing (cancer) cells. Recipients of chemotherapy similarly experience bone marrow suppression and become neutropenic (meaning low in neutrophil counts, similar to the word anemic). They also become thrombocytopenic (low platelet count) and are susceptible to bleeding disorders. Neutropenic cancer patients are treated with a class of drugs called colony-stimulating factors (CSFs).

Colony-stimulating factors are cytokines, hormone-like protein or glycoprotein (sugar-protein) complexes that stimulate a specific cell class or classes to grow, stop growing, or change function. In general, cytokines work by stimulating bone marrow stem cells to proliferate and differentiate into a variety of mature blood cell types. When treated with neutrophil stem cell cytokines, a patient's white blood cell count does not drop as low and does not stay low as long as it would have if the patient had gone untreated. Therefore, for people exposed to moderate or high doses of radiation whose bone marrow is affected, cytokines are an important first line of therapy that can improve the odds of survival.

Neupogen® Production of the neutrophil cytokine naturally present in the body has been genetically engineered by scientists and is marketed worldwide under the trade name Neupogen. This drug (generic name, filgrastim) is FDA approved as adjunctive therapy for patients undergoing aggressive chemotherapy or bone marrow transplants. It can be found in nearly every major hospital in the United States and Europe that has a cancer therapy clinic.

The FDA has not, however, formally approved the drug for treating radiation injury. Even though this drug has already been used in dozens of cases involving victims of high-dose radiation exposure, such use is regarded as "off-label" and takes place at the discretion of attending physicians. Nevertheless, the U.S. government has purchased Neupogen for the Strategic National Stockpile specifically for use in treating patients exposed to medium and high doses of radiation. It is expected that if a national emergency necessitated administering the drug to large numbers of patients, the FDA would issue an emergency use authorization (at the direction of the secretary of Health and Human Services), making this drug available to public health officials and health care providers.

Neupogen can be administered intravenously (IV) or by subcutaneous (SC; under the skin) injections on a daily basis for approximately 2 weeks or until the patient has recovered from neutropenia. Administration of Neupogen in a mass casualty situation, such as a nuclear detonation, will present challenging logistical issues.

Although this drug is generally safe, it may have serious side effects. Daily treatment can lead to enlargement of the spleen and a number of fatal ruptures of this organ have been reported. Some patients have shown severe allergic reactions to this drug, but have responded well to conventional therapies such as antihistamines and steroids. The most common adverse effect is bone pain, seen in approximately one-quarter of all patients receiving the drug. It is best to treat only patients who can benefit from this treatment (Chapters 6 and 11) and to follow up and monitor those receiving this treatment.

One limitation of Neupogen is that its effectiveness in stimulating production of other types of blood cells, such as platelets, is minimal. It is relatively expensive, and the drug should be administered within a day or two of the radiation exposure to be most effective.

Other Examples The drug Neulasta® is the long-acting form of Neupogen. Many cancer patients may be familiar with this drug, which is approved by the FDA for treating patients with low neutrophil counts undergoing cancer therapy. Again, use of this drug to treat radiation-related neutropenia would be at the discretion of the prescribing health care provider. A single administration of Neulasta is significantly more expensive than a single dose of Neupogen. In a mass casualty situation in which eligibility of patients receiving the treatment may not be known with certainty, it may be prudent to treat with Neupogen because daily treatment can be stopped if subsequent evaluation indicates that it is not required.

The search for better drugs continues. An ideal drug for a mass casualty situation (e.g., nuclear detonation) should be effective even if given days after exposure. It should also have a long shelf life, be easy to administer, impart minimal side effects if given to individuals who turn out not to need the treatment, and be relatively inexpensive. Although there have been some promising candidates (e.g., Neumune™), the search for a drug with these traits continues.

The following are details about manufacturing of cytokines and related terminology. (Some readers may choose to skip ahead to the next section.)

Did you know? Filgrastim is made using recombinant DNA technology, meaning that the drug is mass produced by *Escherichia coli* bacteria into which a human gene has been inserted.

Did you know? The generic name for Neulasta is pegfilgrastim. The "peg" in pegfilgrastim refers to a polyethylene glycol unit that is added during the manufacturing of filgrastim protein (or Neupogen). PEG prevents the protein from being broken down in the body, making it longer acting.

Did you know? These naturally occurring cytokines are called "colony-stimulating factors" because of how they were first isolated and discovered. Hematopoietic (blood-forming) stem cells are cultured in the laboratory into small groups or colonies. The substance being tested is added to the cell cultures to see which types (if any) of colonies (or group of stem cells) are stimulated into growth. For example, a substance that stimulates formation of colonies of granulocytes (a category of white blood cells) is called a granulocyte colony-stimulating factor (G-CSF). Filgrastim is a G-CSF. Neutrophil granulocytes (or neutrophils) are the most abundant type of white blood cell.

DRUGS TO TREAT EXTERNAL CONTAMINATION

Drugs are not needed to treat external contamination with radioactive materials. As described in Chapter 9, the most effective way to address external contamination is removal of contaminated clothing, washing with mild soap and tepid water, and shampooing (without conditioner) if possible. Excessive forceful scrubbing is not necessary and should be avoided. In extremely rare cases, certain dilute chemical solutions may be used to clean residual or persistent skin or hair contamination.

DRUGS TO TREAT INTERNAL CONTAMINATION

Radioactive materials can make their way inside the body during breathing, eating, or drinking; by absorption through the skin; or through an open wound. This phenomenon is referred to as internal contamination (Chapter 3). The two most common pathways of internal contamination are via inhalation and ingestion. Once inside the body, radioactive materials may become uniformly distributed in the soft tissues (e.g., muscles) or may concentrate in a particular tissue (e.g., bone, thyroid, or kidneys). Over time, a percentage of the material is eliminated from the body through urine, feces, and sweat. The rate at which radioactive material distributes in the body, is absorbed by a particular tissue, or leaves the body through excretion depends on a number of factors. These include:

- the physical properties of the radioactive element itself (e.g., cesium, plutonium, or others);
- the chemical and physical forms of the material as it enters the body (e.g., nitrate or oxide; size of particles inhaled);
- the route of intake (e.g., inhalation or ingestion); and
- the inherent metabolism of the individual.

As a result, the medical or pharmaceutical approach to address internal contamination varies depending on the specifics of the contamination. The approach suitable for one case may not work and may even be counterproductive in another scenario.

Treatment Objectives and General Approaches

The objective of every treatment for internal contamination is to reduce the time that radioactive materials stay in the body and, if possible, to prevent their absorption into body tissues. The faster the materials are eliminated from the body, the greater is the reduction in future health risks.

A few general procedures may be considered if medically warranted. A number of these procedures may only be useful in occupational accidents when the contaminating event is immediately known and the number of affected individuals is small:

- Gastric lavage or "pumping the stomach" is used to clean out the contents of stomach. It may be warranted shortly (within an hour) after radioactive material has been ingested.
- Pulmonary lavage has been used in occupational environments when workers may accidentally inhale plutonium oxide.
- Emetics are used to induce vomiting shortly after radioactive materials have been ingested.
- Laxatives or enemas are used to empty the intestinal contents.
- Diuretics are used to promote excretion of urine. Drinking beer, for example, is effective in accelerating excretion of tritium from the body. Following consumption of beer, reporting back to work is not recommended! Water, tea, and coffee are also effective at inducing diuresis without the intoxicating side effects.

These types of treatments do not apply to radiation emergencies in which large numbers of people are potentially contaminated, and the amount of internal contamination for most people is expected to be small and not life threatening. A number of radiation drugs can be used in such emergency situations involving internally contaminated populations.

Currently Available Drugs

In this section, three commercially available drugs approved by the FDA for the explicit purpose of treating internal contamination with radioactive materials are described in detail. Many countries and many states within the United States have their own caches of one or more of these drugs:

1. potassium iodide for treating internal contamination with radioactive iodine;
2. Prussian blue for treating internal contamination with radioactive cesium and thallium; and
3. DTPA for treating internal contamination with radioactive americium, plutonium, and curium.

The recommended usage and the limitations of each of these drugs are described in this section.

Using animal models, other drugs have been tested as decorporating agents to help the body get rid of other types of radionuclides. A number of them have been used to treat people in accidental cases of internal contamination. References are provided for additional information on drugs for which minimal human safety data are available.

As you review the drug information in this section, keep in mind that drugs for treating internal contamination share these limitations:

- They cannot protect against external radiation exposure.
- They cannot prevent radioactive material from entering the body.
- They are effective against certain radionuclides, but ineffective against others.
- They are not immediate in their action and do not completely or instantly remove contamination from the body.

> *Terminology:* In discussing internal contamination, the words "intake" and "uptake," which sound similar but have different meanings, may be used. Intake means the radioactive material has made its way into the body through ingestion, inhalation, or absorption through skin or open wound. Uptake means that the radioactive material has been absorbed by an organ or by tissues (e.g., bone, thyroid gland, or muscle). Thus, a quantifiable intake of radioactive material is possible, but only a percentage of that material ends up as uptake in body tissues. Radioactive contamination that is ingested, passes through the digestive tract without being absorbed, and is subsequently excreted is part of the intake but not the uptake.

Potassium Iodide Potassium iodide (KI) is a drug used to protect the thyroid gland from radioactive iodine uptake. It is one of the simplest drugs to understand. What KI can do and its limitations are clear, or at least they should be. However, in the United States, use of KI has on occasion become an emotional, sometimes politically charged subject. Rhetoric about KI has caused unnecessary confusion for people who just want to know the facts.

The thyroid gland needs iodine in the daily diet to function properly and produce metabolic hormones. The thyroid cannot distinguish between stable nonradioactive iodine and radioactive iodine (commonly called radioiodine). Therefore, if radioiodine is present in the body, the thyroid will absorb it in the same way as it would stable iodine. KI is a salt tablet. The active ingredient is stable iodine. When the tablet is taken, the stable iodine quickly (and temporarily) saturates the thyroid. Temporarily saturating the thyroid with stable iodine protects the gland against uptake of any radioiodine that may enter the body. Therefore, KI does not prevent the intake of radioiodine, but it does prevent or reduce its uptake in the thyroid gland.

Finding the target organ saturated, the radioiodine will then be excreted from the body, and the overall radiation dose, especially to the thyroid gland, is substantially reduced. Conversely, large radioiodine uptake by the thyroid gland may cause adverse affects, especially in children. The blocking action of KI protects the thyroid gland from harmful effects of radioiodine, including hypothyroidism, thyroiditis (inflammation of thyroid), and thyroid cancer. This protection is most critical for those under the age of 18, who are at a higher risk of developing thyroid cancer, and for pregnant women, who can pass the radioactive iodine to the fetus.

KI is an FDA-approved drug. It does not require a prescription and it is inexpensive. KI can be mail-ordered from a number of retail pharmacy Web sites and is occasionally found over the counter in some pharmacies. In the United States, KI tablets sold under the brand names Iosat™ and ThyroSafe™ are approved by the FDA (Figures 12.1 and 12.2). ThyroShield™—KI in liquid form—is also approved by the FDA. For children and others who are unable to swallow tablets, ThyroShield offers an alternative, effective formulation for protection of the thyroid gland (Figure 12.3).

In a number of other countries, potassium iodate (KIO_3), which has an efficacy similar to that of potassium iodide, is marketed and used in emergency planning. The recommended dosage of the drug measured in milligrams (mg), however, differs from that of potassium iodide. For example, the recommended daily KI dosage for adults is 130 mg, of which 100 mg is the active ingredient, iodine. In order to get the same amount of iodine, it would be necessary to take 170 mg of potassium iodate. This is simply because of

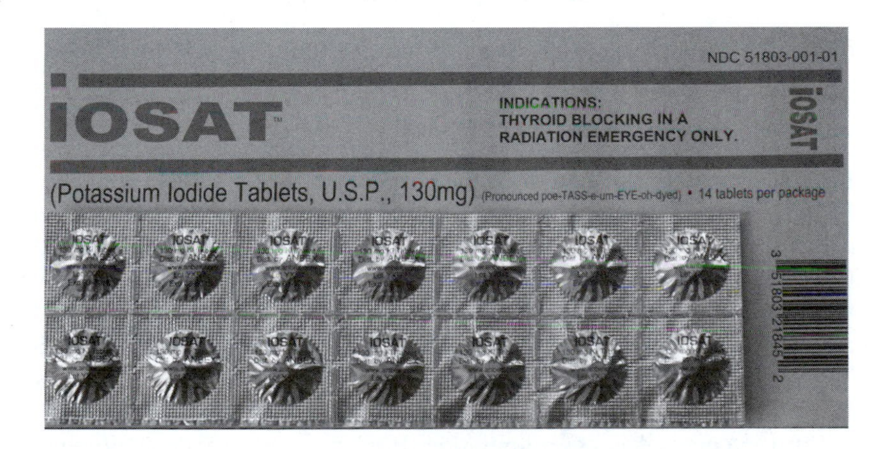

Figure 12.1

Iosat™ is a potassium iodide (KI) tablet approved by the U.S. Food and Drug Administration for protection of the thyroid gland against uptake of radioactive iodine. Each Iosat tablet contains 130 mg of KI, the recommended adult dosage. (Photo courtesy of Anbex.)

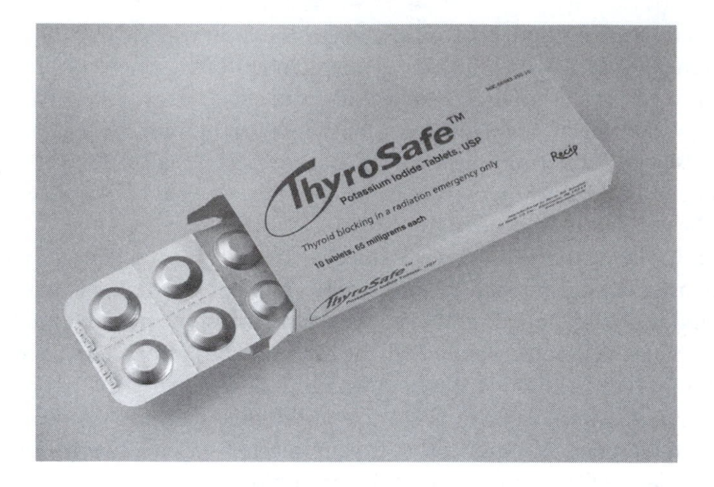

Figure 12.2

ThyroSafe™ is a potassium iodide (KI) tablet approved by the U.S. Food and Drug Administration for protection of the thyroid gland against uptake of radioactive iodine. Each ThyroSafe tablet contains 65 mg of KI, the recommended dosage for children between 3 and 18 years of age weighing less than 70 kg (154 pounds). The recommended dosage for adults is two tablets or 130 mg of KI. (Photo courtesy of Recipharm Inc., Honey Brook, Pennsylvania.)

different molecular weights of the two chemical formulations. Any KI dosage multiplied by 1.3 yields the equivalent dose of KIO_3.

Limitations The effectiveness of KI is time dependent; it is most effective if taken before the radioiodine enters the body or within 3 or 4 hours of intake. KI can be taken even a day before coming into contact with radioiodine and be equally effective. However, if radioiodine has already been inhaled or ingested, there is only a short window of time when KI can be effective at protecting the thyroid gland. At 8 hours after intake, KI has lost more than half its effectiveness. At 24 hours after intake, KI has no effect at all because the thyroid gland has already become saturated with radioiodine. In persons with underlying iodine deficiency, the amount of radioiodine uptake is even greater.

This time dependency is a limitation. If KI cannot be taken as prophylaxis, the primary means of protection, such as sheltering, evacuation, and avoiding a radioiodine-contaminated diet (primarily by not drinking milk produced by animals grazing in the area surrounding the radiation event), become even more important. Even when it is available, KI is always a secondary rather than a primary protective measure.

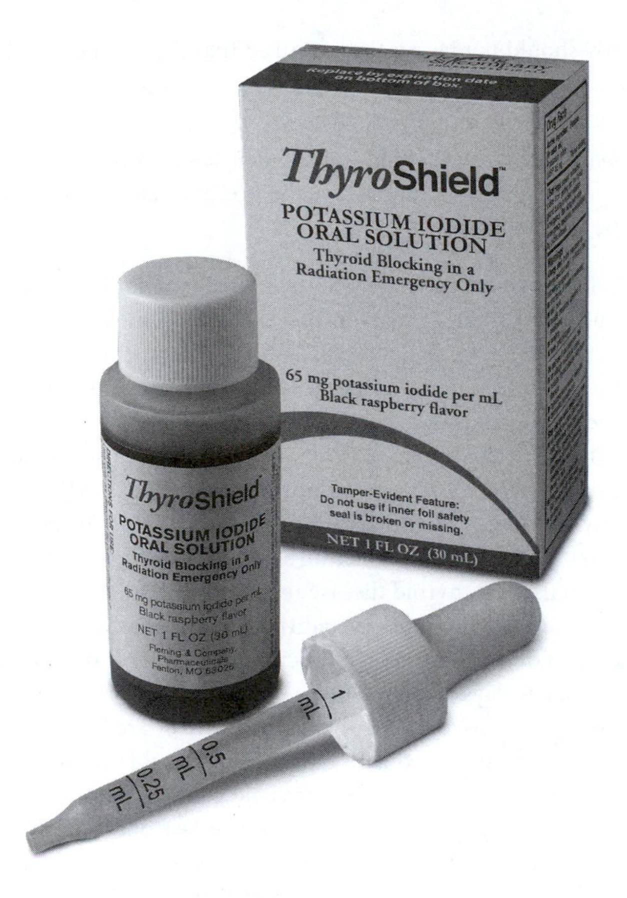

Figure 12.3
ThyroShield™ is a potassium iodide (KI) formulation approved by the U.S. Food and Drug Administration for protection of the thyroid gland against uptake of radioactive iodine. This flavored solution offers an alternative for children and others who are unable to swallow tablets. (Photo courtesy of Fleming Pharmaceuticals.)

An even more important limitation is that KI offers no protection against other types of radioactive materials that may be present. It also does not protect against external radiation levels, which are likely to be present in the environment immediately after a radiation emergency. Therefore, in any radiation emergency, it is critical to follow the primary protective actions recommended by health authorities (such as sheltering or evacuation). Those who ignore these recommendations trying to obtain KI tablets place themselves and their families at risk. On the other hand, people who take

KI promptly should not be falsely reassured that they do not need to follow recommended primary protective actions.

Precautions Doses of KI need to be taken once daily as long as there is potential for intake of radioiodine. Special precautions are necessary for certain groups of individuals:

- Individuals with known sensitivity or allergy to iodine should avoid KI and rely on the primary protective action measures described earlier. Note that a seafood or shellfish allergy does not necessarily mean sensitivity to iodine. Anyone unsure about sensitivity to iodine should consult a doctor.
- Individuals with certain skin disorders (dermatitis herpetiformis or urticaria vasculitis—rare conditions associated with hypersensitivity to iodine) should also refrain from taking KI and rely on primary protective action measures.
- Individuals with thyroid disease (e.g., multinodular goiter, Graves' disease, or autoimmune thyroiditis) should take KI with caution and seek advice from their physician, especially if dosing extends beyond a few days.
- Pregnant women (regardless of age) who are or are likely to become internally contaminated with radioiodine should take KI to protect their own thyroid gland and the thyroid gland of the fetus. The FDA recommends that pregnant women only take a one-time dose of KI, unless primary protection measures (evacuation, sheltering, and control of the food supply) are unavailable and the risk of radioiodine intake is still present. Repeat dosing for pregnant women should be on the advice of public health authorities.
- Radioiodine is readily passed to nursing infants through breast milk. Breast-feeding women who are or are likely to become contaminated with radioiodine should take KI to protect their own thyroid gland. However, any KI passed into breast milk is insufficient to protect nursing infants. The Centers for Disease Control and Prevention (CDC) recommends that nursing mothers contaminated or at risk of becoming contaminated stop breast-feeding and switch to using baby formula or other food. However, if breast milk is the only food available for an infant, nursing should continue. In either case, nursing mothers should be treated with KI. Infants not switched to formula or other food whose mothers take repeat doses of KI, require monitoring for hypothyroidism.

What Type of Radiation Emergency Releases Radioactive Iodine? Where does radioiodine come from and when can it present a health hazard? Radioiodine is a fission product; wherever there is nuclear fission, radioiodine is produced. Accidental release of radioactivity from a nuclear power plant includes radioiodine contamination. Fallout from a nuclear detonation also contains radioiodine. The radionuclide of primary concern is iodine-131 (I-131). It emits beta and gamma radiation and has a half-life of 8 days.

Produced in nuclear reactors for commercial purposes, I-131 is used routinely in thyroid ablation therapy. Veterinarians also use I-131 to treat thyroid disease in cats and dogs. Can I-131 be obtained and used in a dirty bomb? This is possible, but because of its short half-life, it is an unlikely choice for use in a dirty bomb. Therefore, even though a dirty bomb laced with I-131 cannot be ruled out, the more likely radioiodine hazard would be in a case of (1) a serious accident at a nuclear power plant, or (2) fallout following a nuclear bomb detonation.

When either of the preceding incidents occurs, individuals who have not sheltered or evacuated may inhale the radioiodine while it is still in the air. Radioiodine in food and drink serves as another important route by which internal contamination may occur. As noted earlier, radioiodine gets concentrated in milk as dairy farm animals ingest contaminated grass. Goat's milk is an even more concentrated source of radioactive iodine than cow's milk.

Children who drink contaminated milk are at an increased risk of developing thyroid cancer and other thyroid diseases. Unfortunately, this was the case for thousands of young people living in areas surrounding the Chernobyl nuclear power plant. Given proper and timely notification, along with appropriate instructions from public health authorities, such risks can be readily avoided, even in the absence of KI.

Distribution or Predistribution of KI to the Public The International Atomic Energy Agency (IAEA) and the World Health Organization (WHO) have published general recommendations for countermeasures in the case of a nuclear emergency. However, KI distribution plans differ from country to country. In the United States, the Nuclear Regulatory Commission (NRC) and the Federal Emergency Management Agency (FEMA) provide guidance on the issue, and each state with an operating nuclear power plant has its own specific plans for using KI in such an emergency. In some communities in the United States and Europe, KI tablets are predistributed to the population living in the 16-kilometer (10-mile) radius of nuclear power plants.

The effectiveness of this predistribution approach has been questioned. In other similar communities, the authorities have plans (and a stockpile of KI) in place to distribute the pills only when or if it is needed. In other nuclear power plant communities, KI distribution is considered only for emergency workers and not for the general population. In these communities, the emphasis is placed on primary protective action measures, sheltering, evacuation, and control of food supply.

Which approach is better? This question has a tendency to evoke emotional and political reactions. After a thorough investigation of KI policies and distribution experiences in the United States and several countries around the world, the U.S. National Academy of Sciences concluded that each community should decide the best approach after considering the options and specifics of that community. If you live within a 16-kilometer (10-mile) radius of a nuclear power plant, you should know what KI distribution plans (if any) are in place for your community.

As mentioned earlier, KI is inexpensive and the FDA-approved formulation is available without a prescription for anyone who wants to keep a personal supply.

> *Caution:* In any radiation emergency, there may be hazards more serious than radioiodine. The discussion here was for the purpose of presenting how or when KI can be useful, rather than for placing undue emphasis on the relative hazard of radioiodine in radiation emergencies. The availability of KI should not provide a false sense of security to the population at risk, and its unavailability should not cause misplaced priorities in an emergency.

Prussian Blue Prussian blue (PB) is the drug of choice in cases of internal contamination with radioactive cesium and has been used since the 1960s for this purpose. The most well-known application of PB was in the treatment of internally contaminated patients of the 1987 Goiânia accident in Brazil (Chapter 4). Prussian blue is also effective against radioactive thallium and nonradioactive thallium (once an ingredient in rat poison).

PB is taken orally in the form of a capsule three times a day. One reason PB is safe and has few side effects is that it remains in the digestive tract until it leaves the body; it is never absorbed by the body. While in the intestine, PB works like a trap. It binds to any cesium that is already in the intestine or that enters the intestine, keeping it from being reabsorbed by the body. When PB leaves the body through bowel movements, it takes the radioactive cesium with it. The net effect of PB is that it significantly reduces the average time radioactive cesium remains in the body, thereby reducing the time it has to irradiate the body.

The radionuclide of concern is cesium-137 (Cs-137), which has a half-life of 30 years. Together with its decay product, barium-137, Cs-137 emits beta and gamma radiation. The biological half-life of Cs-137 in the human body is approximately 110 days; with PB treatment, this time is shortened to approximately 30 days. The biological half-life (see Chapter 3 for definition) varies

Figure 12.4

Radiogardase™ is a Prussian blue capsule approved by the U.S. Food and Drug Administration for treating internal contamination with cesium and thallium. (Photo courtesy of Heyltex.)

depending on individual metabolism; it is usually shorter in women and still shorter in children. Whatever the biological half-life is for an individual, treatment with PB helps to reduce it by one-half or two-thirds.

> *Did you know?* The process by which any material is (re)absorbed from the intestine and then carried to the liver through the bloodstream is called enterohepatic circulation. The liver filters the material from the blood and passes it into the intestine as a component of bile. Prussian blue interrupts this process by trapping radioactive cesium in the gut, thereby preventing its passage through the intestinal wall into the bloodstream.

Prussian blue, under the trade name Radiogardase™, is approved by the FDA for treating internal contamination with cesium and thallium (Figure 12.4). The U.S. Strategic National Stockpile stores caches of Radiogardase for use in emergencies. The U.S. Department of Energy's Radiation Emergency Assistance Center/Training Site (REAC/TS) and a number of its partners also store some amounts of this drug. A number of state and local jurisdictions store their own supplies as well. Unlike the case with KI, Radiogardase is not an over-the-counter (OTC) medication; it is only available by prescription. Radiogardase is manufactured in Germany and is available worldwide.

> *Did you know?* "Prussian" blue got its name because it was the dye used to color Prussian military uniforms. Accidentally produced in 1704 in Berlin (it was once called Berlin blue), the dye pigment was the earliest of synthetic dyes. It is now widely used in artists' paintings and has wide application as a pigment in research and industry. Only the FDA-approved formulation, however, should be used for medicinal purposes. The chemical name for Radiogardase is ferric hexacyanoferrate(II) with the chemical formula $Fe_4[Fe(CN_6)]_3$. Also see "Precautions."

Limitations The efficacy of PB is not as time dependent as that of KI or, as you will read in the next section, DTPA. As long as the body contains significant amounts of radioactive cesium, PB will be effective in accelerating its removal from the body. The earlier the treatment starts, the better. Treatment can continue for 30 days depending on the extent of contamination.

When Prussian blue was approved by the FDA in 2003, it was touted in the headlines as a "dirty bomb pill." Again, PB offers no protection against external sources of radiation, including a cesium-contaminated environment. It offers no protection to people who enter high-radiation areas or against internal contamination with any radionuclide other than cesium or thallium. For example, it has no effect on internal contamination with iodine, americium, plutonium, uranium, strontium, or a variety of other radionuclides. Even with Cs-137, PB does not "erase" the adverse effects of contamination, although it can reduce them.

Furthermore, the radioactive contamination in the body has to reach a medically significant level before treatment with PB is warranted. In the case of a dirty bomb scenario, large numbers of people may have residual amounts of internal contamination. In cases of residual contamination with cesium, PB treatment is not necessary.

Precautions Prussian blue is safe for adults, including pregnant women, and children aged 2–12 years. Because of insufficient safety data for children under the age of 2, FDA approval of PB in 2003 did not include this particular age group. However, in case of an emergency, the FDA may authorize pediatric use of Prussian blue under an emergency use authorization.

The most common side effects of Prussian blue are upset stomach and constipation. These can easily be treated by coincident administration of other commonly used medications. It is normal for people taking PB to see the color of their stool turn blue during treatment.

Some additional precautions include:

- The CDC recommends that nursing mothers contaminated or at risk of becoming contaminated stop breast-feeding and switch to using baby formula or other food. However, if breast milk is the only food available for an infant, nursing should continue. In either case, nursing mothers should be treated with PB. Scientists and doctors do not believe that PB can pass into breast milk. Mothers who continue to nurse should seek immediate consultation with a physician.
- Individuals who are constipated, have blockages in the intestines, or chronic stomach problems should seek medical advice before taking PB.

- Individuals who have problems swallowing pills or capsules can take Prussian blue by breaking the capsules and mixing the contents in food or liquid. This is equally effective, but it will cause people's mouths and teeth to turn blue during treatment.
- Prussian blue as a dye, paint, or pigment is commonly available from art supply stores and other sources, including the Internet. People should *not* take these products to treat internal contamination or for any other medicinal purpose. Entrepreneurs may also market their formulation of PB for treating radioactive contamination. Only the FDA-approved Prussian blue product should be taken for this purpose.

What Type of Radiation Emergency Releases Radioactive Cesium? Cesium-137 is a fission product and, as with I-131, wherever there is nuclear fission, some amount of Cs-137 will be present. Therefore, releases from a nuclear power plant or a nuclear bomb detonation will contain some amount of radioactive cesium. However, the most likely scenario resulting in people becoming internally contaminated with significant amounts of Cs-137 is an RDD scenario in the form a dirty bomb or by other dispersal methods (see Chapter 5).

DTPA DTPA is a drug used to facilitate the elimination of plutonium, americium, and curium from the body. The drug works by "chelating" or binding to these transuranic elements in the bloodstream to form a complex that is then excreted through urine.

> *Terminology:* "Transuranic" refers to any element heavier than uranium (atomic numbers > 92). All transuranic elements are radioactive. Only plutonium and neptunium are present naturally on Earth in very small, trace quantities. All other transuranic elements are man-made and do not exist naturally. This includes americium-241, which is commonly used in smoke detectors.

Through the years, DTPA has been used to treat hundreds of workers internally contaminated with plutonium, americium, or curium (mainly plutonium). DTPA comes in two chemical forms, a calcium salt (Ca-DTPA) and a zinc salt (Zn-DTPA). The distinction is important and explained in the next section. The recommended method of DTPA administration is by intravenous injection. It can also be administered by nebulized inhalation as an alternative route of administration in individuals whose internal contamination is only by inhalation within the preceding 24 hours. Currently, no approved formulation of DTPA can be administered orally.

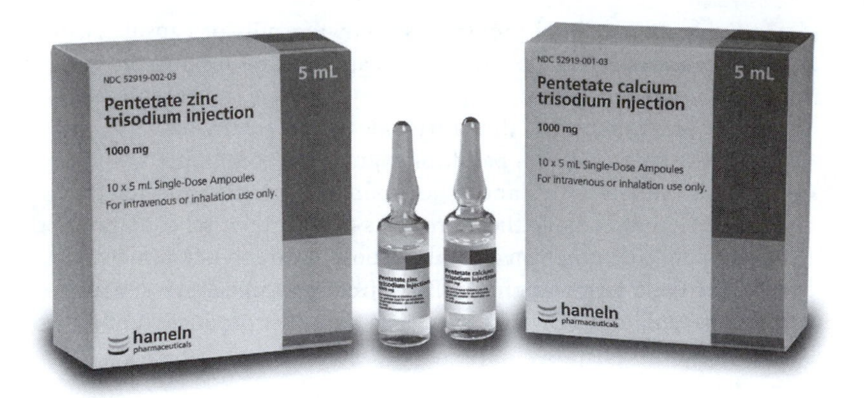

Figure 12.5

Pentetate calcium (or zinc) trisodium, also known as trisodium calcium (or zinc) diethylenetriaminepentaacetate, is commonly referred to as Ca-DTPA and Zn-DTPA. Both are approved by the U.S. Food and Drug Administration for treating internal contamination with plutonium, americium, and curium. (Photo courtesy of Akorn.)

DTPA is most effective if given within the first few hours after contamination. This is when the radionuclides are still in the bloodstream and can be more efficiently chelated and removed from the body. After 24 hours, it becomes more difficult (less efficient) to chelate the radionuclides because uptake by body tissues such as bone, liver, and other organs occurs. Nonetheless, DTPA should still be administered as soon as it becomes available, and at any time it can help decorporate some plutonium, americium, or curium.

The U.S. Strategic National Stockpile stores caches of both Ca-DTPA and Zn-DTPA injection vials for use in emergencies. REAC/TS and a number of its partners also store some amount of this drug. A number of state and local jurisdictions store their own supplies as well. DTPA is manufactured in Germany and distributed worldwide (Figure 12.5).

> *Did you know?* DTPA stands for *d*iethylene-*t*riamine-*p*enta-*a*cetate (hyphens are not used and are only included to aid in pronunciation).

Zinc or Calcium DTPA? The two forms of DTPA have important differences. In the first 24 hours after internal contamination, Ca-DTPA is 10 times more effective in removing the contaminants than is Zn-DTPA. After 24 hours have elapsed, Zn-DTPA and Ca-DTPA exhibit similar chelation efficacy. In the face of repeated administration, however,

Ca-DTPA depletes essential metals from the body, such as zinc, magnesium, or manganese. Assuming equal availability of both products in the first 24 hours after contamination, the FDA recommends that Ca-DTPA be given as the first dose. After 24 hours or for multiple treatments, Zn-DTPA becomes the drug of choice. Multiple treatments are referred to as "maintenance therapy." Zn-DTPA is also the preferred drug for treating pregnant women.

Even though the preceding recommendation is the ideal treatment sequence, it should be noted that if Ca-DTPA is not available or treatment cannot be started within the first 24 hours after contamination, treatment should begin with Zn-DTPA. If Zn-DTPA is not available, Ca-DTPA can be given for continued treatment (maintenance therapy), along with vitamin or mineral supplements that contain zinc. This recommended regimen applies to pregnant women as well.

In many cases, one dose of DTPA may be sufficient, but the length of treatment will depend on an individual's level of internal contamination and his or her other medical conditions (e.g., liver function) as evaluated by a physician.

Limitations DTPA (both calcium and zinc) has been used effectively for decades to treat persons internally contaminated as a result of occupational accidents. Occupational accidents typically occur in controlled environments and usually involve only one or a few individuals. Under these circumstances, workers receive immediate medical monitoring and DTPA can be administered rapidly. The logistics of administering DTPA (by injection) within a few hours of a mass casualty incident will be challenging under anticipated resource-scarce conditions.

It should also be noted that DTPA compounds are most effective in chelating soluble forms of the transuranic metals. If the contaminant is a highly insoluble chemical, the efficacy of chelation is significantly reduced.

Precautions DTPA toxicity is mainly due to chelation and subsequent reduction in serum levels of certain essential minerals such as zinc, manganese, and magnesium. Ca-DTPA causes a more pronounced depletion of essential minerals; coadministration of mineral supplements including zinc can help alleviate this effect. DTPA may cause nausea, vomiting, chills, diarrhea, fever, pruritus (itching), and muscle cramps. A number of medical issues should be considered when DTPA is administered:

- DTPA should be given to individuals with normal renal function. Individuals with serious kidney disease should have their renal function checked before receiving DTPA and then monitored during their treatment course.

- Minors, pregnant women, and individuals with bone marrow suppression should also be closely monitored.
- The CDC recommends that nursing mothers that are contaminated or at risk of becoming internally contaminated stop breast-feeding and switch to using baby formula or other food. However, if breast milk is the only food available for an infant, nursing should continue. In either case, nursing mothers should be treated with DTPA. Any DTPA passed into breast milk is likely insufficient to protect nursing infants. If contamination is suspected, the infants can also be treated with DTPA, with dosage adjusted for weight.
- For individuals suffering from hemochromatosis (iron overload disease), DTPA should be given with caution.
- If nebulized DTPA is used, the drug may aggravate coughing or wheezing in asthmatics.
- The safety and efficacy of nebulized DTPA has not been established for children.
- The safety and efficacy of intramuscular injection of DTPA has also not been established.
- DTPA should not be used to treat persons internally contaminated with uranium or neptunium. The DTPA forms complexes with uranium and neptunium that are not chemically stable and treatment with DTPA may actually result in more adverse effects for these individuals.

In What Type of Radiation Emergency Would DTPA Be Needed? DTPA has already been used to treat individuals who inhaled plutonium as a result of an occupational accident, mistake, or failure of personal protective equipment. Some hypothetical scenarios in which DTPA would need to be administered to a large population include:

- An individual or a terrorist group may obtain sufficient quantities of americium-241 to build a dirty bomb or other radiological dispersal device.
- An individual or a terrorist group may obtain quantities of plutonium for the same purpose. (This is a very unlikely scenario. Plutonium is extremely difficult to acquire; if perpetrators obtain it, they will likely not "waste" the precious plutonium in a dirty bomb.)
- A nuclear weapon accident may occur that does not result in detonation of the weapon, but does release radioactive material into the environment.
- An intentional or unintentional event involving stored nuclear waste occurs, spreading transuranic elements in the environment, and contaminating people.

These scenarios are unlikely to occur, so it is probable that DTPA will not be needed to treat a large, internally contaminated population. Indeed, the logistics of administering DTPA to such a population would be resource and time prohibitive.

Other Drugs

A number of treatments or drugs have been tested in animal studies or used to treat individuals with internal contamination with other radionuclides (Table 12.2). For example, sodium bicarbonate treatment has been used to alkalinize urine and facilitate removal of uranium. Aluminum hydroxide (common antacid) or ammonium chloride can be used to treat contamination with strontium. Dimercaprol may be used to treat contamination with polonium. None of these treatments are FDA approved. Nonetheless, they

Table 12.2 Pharmaceutical Countermeasures for Internal Contamination

Radioelement		**Countermeasure**
Am	Americium	Ca/Zn DTPA[a]
Bk	Berkelium	Ca/Zn DTPA
Cf	Californium	Ca/Zn DTPA
Cs	Cesium	Prussian blue[a]
Co	Cobalt	Penecillamine
Cm	Curium	Ca/Zn DTPA[a]
Au	Gold	Dimercaprol, penecillamine
I	Iodine	Potassium iodide (KI)[a]
P	Phosphorous	Na/K phosphates
Pu	Plutonium	Ca/Zn DTPA[a]
Po	Polonium	Dimercaprol
Ra	Radium	Magnesium sulfate, alginates
Sr	Strontium	Aluminum hydroxide, ammonium chloride, alginates
Tl	Thallium	Prussian blue[a]
H	Tritium	Diuretics
U	Uranium	Sodium bicarbonate
Y	Yttrium	Ca/Zn DTPA

Source: Adapted from NCRP Report 65, 1980. *Management of persons accidentally contaminated with radionuclides.* NCRP Report 161, *Management of persons contaminated with radionuclides,* is expected for publication in late 2009. The FDA-approved drugs, however, are still limited to those listed in this table.

[a] Approved by the U.S. Food and Drug Administration for the radioelement listed.

have shown effectiveness in a number of cases and could be used at the discretion of a competent medical authority. Resources are provided at the end of this chapter to direct interested readers to additional information about these treatments.

FUTURE PRODUCTS

Many products are competing for U.S. government support. Project Bioshield was launched in 2004; it entails a significant commitment to expedite research and development of effective drugs and vaccines to protect against chemical, biological, radiological, and nuclear (CBRN) weapons. As part of that project, eight cooperative centers for medical countermeasures against radiation (CMCRs) were established in 2005. Some new methodologies for rapid field triage and diagnosis of radiation sickness may be developed as a result of this project. However, it remains to be seen what new drugs are developed with research support from this initiative. An oral formulation for DTPA is likely, but as discussed earlier, the need for large-scale treatment of the population with DTPA is questionable.

On a more fundamental basis, development of a generic radioprotective drug remains elusive. Basic research, particularly in the area of gene expression, may present opportunities to come closer to this goal in the future. The ability to treat radiation injury to the hematopoietic system may be improved in the near future. As that happens, radiation injury to other tissues such as the intestine, lung, and kidney will demand development of new treatment approaches. Lastly, as early diagnosis and treatment of cancer become more successful, the long-term health risk of exposure to radiation may become less significant.

HOME REMEDIES AND EXOTICS

There is a market for antiradiation remedies. Nutraceuticals make up a whole class of products that are marketed as dietary supplements. In addition to traditional vitamins, minerals, and proteins, nutraceuticals include herbal and other botanical products, enzymes, and any mixtures of these. Although they are usually presented in medicinal packaging in the form of pills, capsules, or powder, very few are subjected to rigorous safety and efficacy studies or have sufficient scientific evidence to support claims of health benefit. Some of these products claim to have protective effects against radiation.

A few of them have at least a legitimate basis. For example, it has been known for decades that vitamin E has some radioprotective effects. In fact, the Armed Forces Radiobiology Research Institute (AFFRI) has demonstrated this effect in mice. But whether this effect is applicable and extends to people in a radiation emergency is highly questionable. Another example is alginates—salts of alginic acid extracted from brown sea algae. Alginates are effective in inhibiting the absorption of radioactive strontium in the intestine. Some entrepreneurs, however, market seaweed as a product that detoxifies the body following exposure to harmful radiation. This claim is extended to radiation coming from cell phones and computer monitors. Some herbalists recommend servings of miso soup (a traditional Japanese soup) with added seaweed for general protection against radiation.

Other homeopathic approaches include soaking in special detoxification baths, eating certain foods, drinking certain teas, and growing plants around the house that can be used to heal radiation burns. In addition, a number of spiritual healers assert that they can provide remote healing and completely eliminate harmful radiation for their clients. Some healers claim that they emit a field that can significantly reduce levels of ionizing radiation, including alpha, beta, and gamma radiation.

The Internet facilitates this marketing and uninformed individuals are likely to be confused, if not misled.

Did you know? The legendary American author Samuel Clemens (Mark Twain) often experimented in real life with homeopathic approaches. In *A Connecticut Yankee in King Arthur's Court* (1889), he wrote: "Any mummery will cure, if the patient's faith is strong in it." In the case of radiation sickness, his statement will be proven wrong, as he most often was with his other "cures" for illnesses.

Resources

For additional information on FDA-approved medical countermeasures including drug package inserts, see the FDA Center for Drug Evaluation Research (CDER), Drug Preparedness and Response to Bioterrorism page and look under Radiation Emergencies (http://www.fda.gov/cder/drugprepare/default.htm).

For information on other treatments for internal contamination, see the NCRP Report 161 and the review by Goans:

National Council on Radiation Protection and Measurements (NCRP). 2009. Management of persons contaminated with radionuclides. NCRP report no. 161. Bethesda, MD. (This report, expected in late 2009, is an update to NCRP 65 report (1980) and will be published in two volumes, the first of which will be a quick reference guide and the second volume will provide the full documentation for recommended procedures to be used in treating contaminated persons.)

Goans, R. E. 2001. Update on the treatment of internal contamination. In *The medical basis in radiation-accident preparedness*, ed. R. C. Riks, M. E. Berger, and F. M. O'Hara, 201–216. New York: Elsevier.

For a detailed report on KI policy and distribution practices in various states in the United States and several other countries, see:

National Academy of Sciences (NAS). 2004. *Distribution and administration of potassium iodide in the event of a nuclear incident.* Washington, D.C.: National Academies Press.

World Health Organization (WHO). 1999. *Guidelines for iodine prophylaxis following nuclear accidents*, Geneva (http://www.who.int/ionizing_radiation/pub_meet/Iodine_Prophylaxis_guide.pdf).

For a detailed report on use of Prussian blue to treat contaminated patients in Goiânia, Brazil, see Farina, F., C. E. Brandao-Mello, and A. R. Oliveira. 1991. Medical aspects of ^{137}Cs decorporation: The Goiânia radiological accident. *Health Physics* 60:63–66.

Additional related information can be found on the following Web sites:

The U.S. Department of Energy, Radiation Emergency Assistance Center/Training Site (REAC/TS) web site (http://orise.orau.gov/reacts/).

The U.S. Department of Health and Human Services, Radiation Event Medical Management (REMM) Web site (http://www.remm.nlm.gov/int_contamination.htm#blockingagents).

The U.S. Centers for Disease Control and Prevention, Radiation Emergencies Web site (http://emergency.cdc.gov/radiation/).

13

GOVERNMENT RESPONSE TO RADIATION EMERGENCIES

OVERVIEW

Effectively responding to a major radiation emergency requires coordination among many response agencies at all levels of government.* The dynamics of this coordinated response will depend on the local government structure and authority, as well as its relationship to the state and federal government. In the United States, local and state government agencies (including local public health and emergency management agencies) are fully empowered to fulfill their responsibilities and exercise their authority. As such, local governments "own" the emergency and are responsible for protecting the health and safety of the population under their jurisdictions. Response to a major radiation emergency, however, will require extensive resources, beyond what is available locally; support from state and federal governments will be necessary to assist local authorities in meeting the needs of the affected population.

This chapter provides a broad overview of such coordinated government response. Even though some of the specifics may apply only to the United States, the principles involved, the types of resources needed, and the extent of planning required are applicable to any jurisdiction in any country. Extensive planning tools are available from the International Atomic Energy Agency (IAEA) to facilitate development of similar infrastructures in all countries. A number of key resources are identified in this chapter.

* The term "response" in government jargon includes any immediate actions taken to save lives, protect property and the environment, and meet basic human needs. "Response" also means implementing any existing emergency plans and actions taken to support recovery from the incident.

In a nutshell, the local police, fire, and emergency medical services (paramedics and emergency medical technicians) take on the major burden of any emergency in its first critical hours. In the first few days, local emergency management and public health agencies have critical roles in containing and managing the incident, communicating to the public, and providing information and protective action recommendations as necessary. Collaboration and coordination with local and federal law enforcement are essential. As help arrives from neighboring jurisdictions and the federal government, the success of the effort depends on clear chains of command as well as clear and coordinated lines of communication. All of these necessities require advance planning and extensive training.

ALL EMERGENCIES ARE LOCAL

When any emergency occurs, the responders on the scene—firefighters, police, and emergency medical services (EMS) personnel—are all *local* responders. The first hospitals to receive the injured are *local* hospitals. It is also incumbent on *local* public health and emergency management officials to assess the situation and provide information and instructions to the public so that people's health, safety, and property are protected. If a segment of the population needs to be evacuated, it is a city, county, or a state official who has the authority to order that evacuation. It will take hours for outside assistance to begin to arrive and a few days, at best, for the full extent of outside resources to become available.

Many local communities have adopted the "YOYO*-72" principle—meaning that the community is on its own for 72 hours before any outside help arrives. This rule assumes the current emergency is the only major national incident and is receiving the full attention of the country. If multiple incidents have occurred or are expected to occur (as they did on September 11, 2001), or if the incident is expected to affect neighboring states (as in the case of a radioactive plume or nuclear fallout crossing multiple state lines), the full presence of federal responders and assistance from those neighboring states cannot be expected. In fact, outside help may not arrive at all. Under these circumstances, many local communities may be on the "YOYO" status for much longer. In any case, after several months have passed, most outside responders leave. *Local* agencies, again, need to deal with long-term consequences of the incident.

The phrase "all emergencies are local" has become a cliché in the world of emergency management. Everyone acknowledges this reality—at least in

* YOYO is an acronym for "You're on your own."

words. However, when precious resources are allocated for emergency preparedness, local jurisdictions do not typically receive the resources they need to improve their readiness. This situation is especially true in the case of a radiation emergency, which is regarded as a rare occurrence. Local jurisdictions are more likely to spend their limited resources on issues that they feel are more likely to occur. As a result, most local jurisdictions are not as prepared to deal with a radiation emergency as they would like to be. It is paramount that local response agencies receive the necessary training and resources to prepare themselves for such a response. Training and planning for radiation emergency response at the local level is best when it is coordinated with each state's radiation control program. See the "Resources" section at the end of this chapter to find out how to locate the radiation control program in a state.

Tiered Response

One of the key response doctrines in the United States is referred to as "tiered response." It states that "incidents must be managed at the lowest possible jurisdictional level and supported by additional capabilities when needed."* A jurisdiction's chief executive officer (the mayor, city manager, or county manager) is responsible for ensuring the health and safety of the population in that jurisdiction. Local authorities should establish strong working relationships with each other and with private sector and nongovernmental organizations in their communities.

If a nuclear incident occurs anywhere in the country, millions of people may be displaced and every community in the country is likely to be affected as it receives a portion of this large population needing shelter and health and medical assistance. A close working relationship among all local jurisdictions and efficient use of available private sector and volunteer resources can make a big difference in meeting the needs of the community. In Chapter 15, the topic of community resiliency is discussed in detail.

NATIONAL RESPONSE

In a true disaster, by definition, existing resources are not adequate to meet the needs of the situation. Most natural disasters (earthquakes, hurricanes, large fires) and certainly a terrorism incident in a given jurisdiction fall into this category and require assistance from neighboring jurisdictions, the federal government, and nongovernmental organizations and volunteers. The

---

* The National Response Framework core document is available from the NRF Resource Center (www.fema.gov/emergency/nrf/index.htm).

coordination of this effort requires carefully thought out plans. Naturally, the larger the size of the government is and the more diverse and abundant outside resources are, the more challenging it is to coordinate these support efforts.

Another factor at play is the bureaucratic structure. In countries with a more or less vertical hierarchy of authority, the coordination effort is simpler and, in some instances, more efficient. In the United States, the civilian bureaucracy has a more horizontal structure. This feature has the distinct advantage of empowering various government entities and nongovernmental organizations in a positive way, but it does make it more challenging to coordinate the vast national response resources.

It is not the intent here to delve into the alphabet soup of U.S. government resources, teams, concepts, and terminology—all of which can be found elsewhere and are likely to be changed or revised anyway. Instead, the response structure in the United States will be broadly reviewed and references for more detailed information will be provided in the "Resources" section. Along the way, a few acronyms will be presented and a few resources specific to radiation emergencies will be listed. Also, international resources available to help develop national radiation emergency response plans where they may be lacking will be described.

As you read this material, it is important to note the following:

- The state is the gateway to federal government assistance. When an incident grows beyond the capability of a local jurisdiction, the local emergency manager should contact the state, which may then ask for federal assistance.
- Federal government resources, even under the best of circumstances, are still limited.
- It takes time, perhaps days, for some outside resources to arrive.
- It is vital for local jurisdictions to be prepared to manage the incident on their own, at least initially.
- It is essential to establish emergency response plans for integrating the incoming federal and state assets with local resources for a unified approach to managing an emergency.
- If emergency response plans are not practiced and tested regularly, they offer very little value.

The Framework

In the United States, a document called the National Response Framework (NRF) provides the guiding principles that enable all response partners to contribute to a unified national response to disasters and emergencies—from the smallest incident to the largest catastrophe. The NRF defines key principles, structures, and the roles of all agencies and partners, and it provides a single, comprehensive approach to managing and recovering from a domestic incident. The framework applies to terrorist attacks, major disasters, or any other emergencies; in other words, it is an "all-hazards" approach.

Some History It is useful to know how this all-hazards document, the NRF, evolved into being. The history is described here for interested readers and it is not necessary to commit any of the information to memory. Prior to September 11, 2001, a document called the Federal Response Plan (FRP) provided a process and structure for the coordinated delivery of federal assistance in any major disaster or declared emergency. Other emergency response plans were also in existence. For example, the National Contingency Plan (NCP) served as the federal government's blueprint for responding to oil spills and hazardous substance releases.

In addition, the Federal Emergency Management Agency (FEMA), the Federal Bureau of Investigation (FBI), and five other federal agencies developed the Interagency Domestic Terrorism Concept of Operations Plan to outline how the federal government would respond to a terrorist threat or incident, including one involving weapons of mass destruction. Of more relevance is the fact that the Federal Radiological Emergency Response Plan (FRERP) was written specifically to address radiation emergencies.

The tragic events of September 11, 2001, resulted in fundamental changes to how the U.S. government responds to disasters. The Homeland Security Act of 2002 created the Department of Homeland Security (DHS), a mammoth organization responsible for coordinating federal operations and resources within the United States to prepare for, respond to, and recover from terrorist attacks, major disasters, and other emergencies. Furthermore, the Homeland Security Presidential Directive (HSPD-5) issued in February 2003 called for a single, comprehensive National Incident Management System (NIMS) to enhance the ability of the United States to manage domestic incidents. As a result, the National Response Plan (NRP) was developed in 2004 and went into effect in April 2005. One of the annexes of this plan, the Nuclear/Radiological Incident Annex, superseded the FRERP and became the document that defines agency roles and responsibilities in a radiation emergency.

The NRP received a challenging test just 4 months after going into effect. On August 29, 2005, the world witnessed another tragedy in the United States as Hurricane Katrina made landfall on the southeast coast of Louisiana and devastated the city of New Orleans. Government response to that hurricane received severe criticism. The Post-Katrina Emergency Management Reform Act of 2006 (PKEMRA) resulted in reorganization of DHS and FEMA. The national response "plan" went under major revisions, was renamed the national response "framework," and went into effect in 2008. A revised Nuclear/Radiological Incident Annex remains in effect.*

Organization Through all the head-spinning changes since 2001, what has not changed is that federal government response and support are still organized in 15 "bundles" called emergency support functions (ESFs). As the name implies, the grouping of resources from various agencies of government is done by "functions" and each ESF has a coordinating agency to lead that effort. The coordinating agency is identified on the basis of authorities and resources. For example, public health and medical services is one of the support bundles and is referred to as ESF 8. As one might expect, the U.S. Department of Health and Human Services is the coordinating federal agency for that support function. Table 13.1 lists all 15 emergency support functions and the coordinating agency for each ESF.

The NRF describes the purpose and primary functions of each ESF, designates its coordinating agency, and identifies all support agencies and a listing of what and how each of these agencies contributes to the process. In addition, the NRF includes a number of annexes dealing with specialty topics. As mentioned earlier, one of those annexes is the Nuclear/Radiological Incident Annex, which addresses radiation incidents in particular.

> *Did you know?* Depending on what kind of facility or what type of radioactive material is involved in a radiation emergency, the federal coordinating agency could be any one of the following agencies: Department of Defense (DoD), Department of Energy (DOE), Department of Homeland Security (DHS), Nuclear Regulatory Commission (NRC), Environmental Protection Agency (EPA), National Aeronautics and Space Administration (NASA), U.S. Customs and Border Protection, or the U.S. Coast Guard.
>
> *Did you know?* In case of any deliberate attacks on the United States involving nuclear facilities, radioactive materials, dirty bombs, or nuclear bombs, DHS will serve as the coordinating agency and the Department of Justice (DOJ) will assume law enforcement coordination activities.

What about the Military? Without a doubt, the military can play a critical role in responding to a catastrophic domestic incident. Members of the armed forces are well trained and well equipped, and they offer valuable operational capabilities. These capabilities include a robust communications infrastructure, logistics and planning, search and rescue operations, and

---

* Other specific incident annexes of the NRF include biological incident, catastrophic incident, food and agriculture incident, mass evacuation, cyber incident, and terrorism incident law enforcement and investigation. All are available from www.fema.gov/emergency/NRF/

Table 13.1 Emergency Support Functions (ESFs) in the National Response Framework

ESF	Coordinating Agency	Responsibilities
ESF #1 Transportation	Dept. of Transportation	Aviation/airspace management and control; transportation safety; restoration and recovery of transportation infrastructure; movement restrictions; damage and impact assessment
ESF #2 Communications	Dept. of Homeland Security (National Communications System)	Coordination with telecommunications and information technology industries; restoration and repair of telecommunications infrastructure; protection, restoration, and sustainability of national cyber- and information technology resources; oversight of communications within the federal incident management and response structures
ESF #3 Public works and engineering	Dept. of Defense (U.S. Army Corps of Engineers)	Infrastructure protection and emergency repair; infrastructure restoration; engineering services and construction management; emergency contracting support for life-saving and life-sustaining services
ESF #4 Firefighting	Dept. of Agriculture (U.S. Forest Service)	Coordination of federal firefighting activities; support to wild land, rural, and urban firefighting operations
ESF #5 Emergency management	Dept. of Homeland Security (FEMA)	Coordination of incident management and response efforts; issuance of mission assignments; resource and human capital; incident action planning; financial management
ESF #6 Mass care, emergency assistance, housing, and human services	Dept. of Homeland Security (FEMA)	Mass care; emergency assistance; disaster housing; human services
ESF #7 Logistics management and resource support	General Services Administration and FEMA	Comprehensive, national incident logistics planning, management, and sustainability capability; resource support (facility space, office equipment and supplies, contracting services, etc.)

Continued

Table 13.1 Emergency Support Functions (ESFs) in the National Response Framework (*Continued*)

ESF	Coordinating Agency	Responsibilities
ESF #8 Public health and medical services	Dept. of Health and Human Services	Public health; medical; mental health services; mass fatality management
ESF #9 Search and rescue	Dept. of Homeland Security (FEMA)	Lifesaving assistance; search and rescue operations
ESF #10 Oil and hazardous materials response	Environmental Protection Agency	Oil and hazardous materials (chemical, biological, radiological, etc.) response; environmental short- and long-term cleanup
ESF #11 Agriculture and natural resources	Dept. of Agriculture	Nutrition assistance; animal and plant disease and pest response; food safety and security; natural and cultural resources and historic properties protection; safety and well-being of household pets
ESF #12 Energy	Dept. of Energy	Energy infrastructure assessment, repair, and restoration; energy industry utilities coordination; energy forecast
ESF #13 Public safety and security	Dept. of Justice	Facility and resource security; security planning and technical resource assistance; public safety and security support; support to access, traffic, and crowd control
ESF #14 Long-term community recovery	Dept. of Homeland Security (FEMA)	Social and economic community impact assessment; long-term community recovery assistance to states, tribes, local governments, and the private sector; analysis and review of mitigation program implementation
ESF #15 External affairs	Dept. of Homeland Security	Emergency public information and protective action guidance; media and community relations; congressional and international affairs; tribal and insular affairs

Source: U.S. Department of Homeland Security. 2008. National response framework (available from www.fema.gov/emergency/nrf/).

Figure 13.1

A convoy of military vehicles crossing floodwaters on Interstate 10 headed toward New Orleans in the aftermath of Hurricane Katrina as part of the disaster relief effort, September 11, 2005. (Photo: U.S. Department of Defense.)

security. Their value was clearly demonstrated during response to Hurricane Katrina in 2005. Even though there were delays requesting their assistance, by September 5 (1 week after the hurricane's landfall), military helicopters had performed 963 search and rescue, evacuation, and supply delivery missions (Figures 13.1 and 13.2).

Military personnel also assisted with other needs. For example, two C-130 firefighting aircraft and seven helicopters supported firefighting operations in New Orleans. Military aircraft flew mosquito abatement aerial spraying missions over 2 million acres to prevent the spread of mosquito- and waterborne diseases. In addition to providing much-needed security, military personnel also performed such missions as salvage, sewage restoration, relief worker billeting, air traffic control, and fuel distribution.*

However, there were serious challenges in the use of military in that instance. The limitations under federal law and military policy caused the active duty military to be dependent on requests for assistance, which delayed and slowed the response. Furthermore, because active duty military

* The federal response to Hurricane Katrina: Lessons learned, February 2006 (available from www.whitehouse.gov/reports/katrina-lessons-learned/).

Figure 13.2
An elderly evacuee rescued from a rooftop in New Orleans, Louisiana, by the U.S. Navy, September 5, 2005. (Photo: U.S. Department of Defense.)

serves under the command of the president and the National Guard serves the state governor, challenges in coordinating their operations ensued.

Posse Comitatus The Posse Comitatus Act of 1878 prohibits the direct use of federal military troops in domestic civilian law enforcement, except where expressly authorized by the Constitution or acts of Congress. Direct involvement in law enforcement includes search, seizure, and arrest. In fact, violation of this act is punishable by fine and imprisonment.* This prohibition applies to all uniformed services except the U.S. Coast Guard. The prohibition does not apply to the Army National Guard and the Air National Guard when they are deployed in state active duty status and are under the command of the state governor. In this case, the National Guard forces are said to be under Title-32 active duty status, referring to U.S. Codes. Conversely, Title-10 active duty forces—Army, Navy, Air Force, Marines, and National Guard forces on federal duty—are under the command of the president and the posse comitatus restrictions apply to them.

* Section 1385 of title 18, United States Code.

After Hurricane Katrina, the government realized the need to clarify the circumstances when the Department of Defense will temporarily lead the federal response to a disaster and how the military assets can be used to assist local and state efforts. A nuclear detonation catastrophe is an example of when such assistance from the military will be desperately needed. The U.S. Northern Command (USNORTHCOM) is the military command charged with defending the U.S. homeland. USNORTHCOM is forming a well-trained and well-equipped task force designed to respond to chemical, biological, radiological, nuclear, and explosive (CBRNE) incidents with specialties in rescue, decontamination, and medical support. This force will supplement state-based National Guard Weapons of Mass Destruction Civil Support Teams and the National Guard CBRNE Enhanced Response Forces.

Did you know? The Posse Comitatus Act of 1878 came at the end of the Reconstruction period that followed the American Civil War. It ended the federal troops' occupation of the former Confederate states.

Did you know? Posse comitatus is the Latin phrase for power or authority of the county (not country). Most people have heard the word "posse" in American western films when a sheriff summons a group of people to assist in a law enforcement activity.

Acronyms You Need to Know It is difficult to avoid acronyms when discussing anything related to government. The acronyms that you really need to know or remember obviously depend on your role and responsibilities or where your interests lie. If you have any response roles, you should be familiar with two important acronyms: NIMS and ICS.

The National Incident Management System (NIMS) was mentioned earlier. The importance of NIMS is that it allows first responders and emergency managers from different jurisdictions and disciplines to speak the same terminology and work under the same management structure. Thus, they can work more closely together to respond effectively to any emergency.

An integral part of NIMS is what is referred to as ICS, the Incident Command System. ICS is a standardized incident management concept that can be applied to all types of hazards or emergencies, small or large. It provides a scalable, flexible, and adaptable framework for all responders. All emergency responders in the United States receive training in ICS principles and nomenclature. Free online courses teach the concepts of NIMS and ICS in detail (see "Resources" at the end of the chapter).

Did you know? The incident command system was originally developed in the 1970s when a task force led by the U.S. Forest Service devised it to improve communication and management of resources in fighting multi-jurisdictional wildfires. Implementing ICS methodology for all types of emergencies is encouraged internationally and it is already in use in a number of other countries.

Federal Assets about Which You Should Know The civilian government offers a number of valuable resources, including funds, equipment, personnel, and technical expertise. This section describes a number of assets specific to radiation emergencies. However, keep in mind that these resources are more limited than may be imagined or desired. In a case of multiple and simultaneous incidents, these resources would be severely stretched.

- The Interagency Modeling and Atmospheric Assessment Center (IMAAC) has computer modeling capabilities to predict remotely movement of a radioactive plume or fallout using real-time meteorological data. IMAAC provides these predictions, in real time, to authorities and responders who are managing the incident, making time-sensitive evacuation or stay-in-place decisions.
- The U.S. EPA and the DOE each has regional teams of radiation experts that can be on-site in a matter of hours.
- The Aerial Monitoring System (AMS) includes fixed-wing aircraft (small planes) and rotary-wing aircraft (helicopters) with radiation detection equipment to measure radiation levels over large areas of land or locate lost radiation sources and transmit the data in real time.
- The Federal Radiological Monitoring and Assessment Center (FRMAC) is established shortly after an incident and is responsible for coordinating all environmental radiological monitoring, sampling, and assessment activities. Its teams of experts (from multiple federal agencies) begin to arrive within 24 hours with radiation monitoring equipment; they are fully deployed within 2–3 days to assist local and state responders in coordinating the collection and interpretation of radiation monitoring data.
- The Radiation Emergency Assistance Center/Training Site (REAC/TS), which has expertise in treating radiation exposure victims, provides specialized medical assistance. This assistance is in addition to the National Disaster Medical System (NDMS), which can deploy in all kinds of emergencies where state authorities ask for medical assistance.
- The Advisory Team for Environment, Food, and Health is composed of technical experts from multiple federal agencies who can provide coordinated technical advice on behalf of the federal government to decision makers on radiation matters dealing with environmental contamination, food safety, human health, and animal health.

International

Useful emergency response plans are those that are tailored for a specific jurisdiction, taking into account the local capabilities, requirements, and government bureaucracy. The IAEA provides a number of excellent tools that can be used for this purpose. For example, the 2003 IAEA manual, *Method for Developing Arrangements for Response to a Nuclear or Radiological Emergency,* provides a step-by-step approach and detailed descriptions of what is needed to establish a national radiation emergency response to address a variety of radiation incidents.

MUTUAL AID AGREEMENTS

Whether at the county or state level, every jurisdiction should coordinate with its neighboring jurisdictions to exchange mutual aid and support in case of an emergency. In the United States, such local and regional agreements have existed for some time. For example, the Southern Mutual Radiation Assistance Plan was established in 1973 to provide a mechanism to coordinate radiation emergency assistance among eight southeastern states—primarily for nuclear power plant emergencies. Membership of this particular mutual assistance compact grew to 13 states in 1975 and Virginia joined the group in 1990. The types of resources to which these regional compacts have access are generally limited.

The Emergency Management Assistance Compact (EMAC) was ratified by the U.S. Congress and signed into law in 1996; it offers a mechanism to provide a wide array of mutual aid and assistance between states addressing liability and cost reimbursement issues.* All 50 states, the District of Columbia, Puerto Rico, Guam, and the U.S. Virgin Islands have enacted legislation and are members of EMAC. This mechanism for mutual aid agreement and partnership can address all types of emergencies—hurricanes, earthquakes, wildfires, toxic waste spills, and terrorism incidents.

Although EMAC has been used effectively many times since its inception, the usefulness of these mutual aid agreements should not be overstated. As noted earlier, if a major radiation incident involving a radioactive plume crossing state lines or a large evacuation of displaced population across several state lines occurs, assistance from neighboring or other jurisdictions may not be as readily available.

* Public Law 104-321.

PRIVATE SECTOR AND NONGOVERNMENTAL AGENCIES

It is a fact that government agencies cannot work alone in protecting the lives, health, and safety of citizens. Partnerships with private sector businesses (health care, energy, water, communications, transportation, security, etc.) and nongovernmental agencies (houses of worship, American Red Cross, Citizen Corps, and others) are crucial in protecting and restoring infrastructure, improving the quality of life, and helping the community to recover. These partnerships should be coordinated at all levels of government. The discussion of community preparedness in Chapter 15 goes into more detail about this.

CHALLENGES

Successfully implementing a government response to a major radiation emergency involves overcoming all sorts of challenges. A few of the important ones are highlighted next.

Unified Command

The concept of unified command is another major response doctrine that is an integral part of the ICS. "Unified command" simply means that when an incident is large enough to involve authorities in multiple jurisdictions and/or multiple agencies with different legal, geographic, and functional authorities, each will designate a representative, and these representatives will work together to manage the incident through a single entity called Unified Command. Through this structure, all these multiple jurisdictions and agencies maintain their own authority and accountability, but will work toward meeting mutually developed incident management objectives. To put it in simple terms, *unified command is meant to bring order to the kitchen when there are too many cooks in it.*

Unified command is a great concept, but its implementation in a time frame necessary to manage the incident effectively presents a huge challenge. Furthermore, if the armed forces are assisting with the response, incorporating them under the civilian incident command or as part of the unified command structure may present challenging operational issues.

Communication

Communication is the single most challenging issue in any emergency management situation. There are multiple layers of communication. Certainly, communication issues among first responders, emergency managers, and decision makers in multiple jurisdictions are important, and it is a lesson that has been repeatedly "learned" and continues to be an issue. Here, let us focus instead on communication between the authorities and the public.

Emergency management authorities and public health authorities receive specialized training on how to communicate with the public directly or through the media. Delivering public messages and communicating to the public about health risks in an emergency is a science. Templates called "message maps" are a set of organized statements in plain language and prepared in advance to answer anticipated questions in an emergency. This is the strategy that Mayor Rudolph Giuliani had practiced extensively, and he used it effectively in the immediate aftermath of the September 11 attacks in New York City. These messages usually adhere to a 27-word/9-second/3-message limitation—called the 27/9/3 rule—to help ensure that any spokesperson is quoted accurately by the media.* This rule recommends limiting the public message in this way because

- 27 total words is usually the length of direct quotes in print media;
- 9 seconds is the usual length of a sound bite on radio or television; and
- 3 key messages are all that the public can process during a high-stress situation.

Even though the science of risk communication is advanced, this issue is still likely to present a challenge as multiple response agencies and authorities try to gather information and issue consistent and accurate statements. Furthermore, when the subject is radiation and the word itself creates so much public anxiety, effective risk communication becomes more challenging and even more vital.

Aside from the "science" of communication, the basic qualities of openness, honesty, clarity, timeliness, empathy toward those affected, and credibility of the person communicating will be important.

> *Did you know?* Immediately after the atomic bombing of Hiroshima on August 6, 1945, the Japanese authorities sought to control and even discount the extent of reported injuries. One advisory addressed to Hiroshima residents proclaimed: "All burns should first be bathed in solutions of half

* Hyer, R., and V. Covello. 2007. *Effective media communications during public health emergencies: A WHO handbook*. Geneva, Switzerland: World Health Organization (available from www.who.int/csr/resources/publications/WHO_CDS_2005_31/en/).

> sea water and half fresh water; this way we can fully defend ourselves against this kind of attack."*

Information will flow to the public through the media from a variety of sources, rather than only from the response authorities. Experts will be interviewed; some will be knowledgeable about the situation and many will perhaps not be as well informed. Therefore, the public should expect that not all information they receive will be consistent. The public should primarily look for information and advice coming directly from public health and emergency management authorities and appreciate that it takes time to gather accurate information. Often there is a sense of mistrust when it comes to government: "What are they not telling us?" It is to be hoped that response authorities will benefit from some level of trust from the public in the beginning and be able to maintain that trust throughout the response and recovery period.

Human Factors

Talking about various local and federal resources and integrating and coordinating with them, it is easy to forget that these resources are *people* or are managed by *people.* An attempt can be made to improve emergency response plans and mechanisms to share and coordinate resources to perfection, but it will always come down to the people who serve in various levels of government leadership and people who will be implementing these preparedness and response actions. Are people in key leadership positions competent team players focused on public service as true public servants? Are they more interested in agency turf battles, bureaucratic games, or advancement of their careers? This "human factor" is difficult to measure or predict, but it certainly plays a role in the effectiveness of preparedness efforts at all levels of government, as well as the response to a radiation or any other type of emergency.

Experience

Every community has experience in dealing with some type of emergency, such as winter storms, wildfires, earthquakes, or floods. For example, many

* Ishikawa, E., and D. L. Swain, trans. 1981. *Hiroshima and Nagasaki: The physical, medical, and social effects of the atomic bombings*, p. 496. New York: Basic Books. This information is quoted from historical documents published by the Hiroshima Prefectural Office. 1972. *Hiroshima Kenshi [History of Hiroshima Prefecture], A-bomb data volume.* Hiroshimaken, Hiroshima, p. 116.

communities along the Gulf or Atlantic Coasts in the United States have a wealth of experience dealing with hurricanes. But when a Category 3 or 4 storm hits these coasts, even with several days of warning and anticipation, many challenging emergency response issues can still be expected.

The fact is that it is fortunate that a nuclear or radiological terrorism incident has never occurred in this country. However, if such an incident were to take place, it would be a first and a flawless government performance should not be expected. Even though a nuclear emergency is a low-probability event, it is a high-consequence catastrophe for which people need to train and prepare so that its impact can be lessened.

Resources

The National Response Framework (NRF) Resource Center is an online reference that provides access to NRF and all supporting documents (www.fema.gov/emergency/nrf/).

The Conference of Radiation Control Program Directors (CRCPD) is made up of radiation control programs in each of the 50 states. You can find the contact information for the radiation control authority in your state on this organization's Web site (www.crcpd.org).

The following CRCPD report contains educational and operational information that is useful for professionals developing radiation emergency response plans as well as responders to a radiation emergency:

The Conference of Radiation Control Program Directors (CRCPD). 2006. *Handbook for responding to a radiological dispersal device: First responder's guide—The first 12 hours.* Frankfurt, KY: CRCPD.

The following guidance issued by the Homeland Security Council provides local and state planners with recommendation on how to prepare and respond to a nuclear detonation incident before any federal assets become available:

Homeland Security Council. 2009. Planning guidance for response to a nuclear detonation: First edition. Available from www.hsdl.org

The following documents from the IAEA provide detailed information on how to develop a national radiation emergency response plan:

IAEA. 2003. *Method for developing arrangements for response to a nuclear or radiological emergency.* Vienna: IAEA (accessible from www-pub.iaea.org/MTCD/publications/PDF/Method2003_web.pdf).

IAEA. 2002. *Preparedness and response for a nuclear or radiological emergency.* IAEA Safety Standard Series GS-R-2. Vienna: IAEA (www-pub.iaea.org/MTCD/publications/PDF/Pub1133_scr.pdf).

The following NCRP report provides valuable information regarding consequence management of radiation emergencies, command and control, public communication, planning and qualifications for responders, and a number of appendices useful for developing emergency response plans: NCRP. 2001. *Management of terrorist events involving radioactive material.* NCRP report no. 138. Bethesda, MD.

Information about the Emergency Management Assistance Compact can be found at www.emacweb.org

The Federal Emergency Management Agency (FEMA) Emergency Management Institute offers free online courses on a variety of topics, including the ICS (http://training.fema.gov/IS/crslist.asp).

This EPA Web site provides a 40-minute instructional video on the topic of message mapping delivered by Dr. Vincent Covello of the Center for Risk Communication (http://www.epa.gov/nhsrc/video/MMap112806.wmv).

This EPA guide is a crisis communication resource for emergency responders and federal, state, and local officials communicating with the public and the media during a radiological crisis: U.S. Environmental Protection Agency. 2007. *Communicating radiation risks: Crisis communication for emergency responders.* EPA-402-F-07-008. This resource is not available online.

14

RESPONDING AS PROFESSIONALS

OVERVIEW

Responders to a radiation emergency include a wide spectrum of professionals, such as law enforcement, firefighters, emergency medical services, hazardous material (HazMat) teams, hospital staff caring for victims as they arrive, public health and emergency management professionals, public information officers, radiation professionals trained specifically to deal with such emergencies, and many others. In a broader context, professionals working in vital service industries who report back to work, when it is safe to do so, are also "responding" to the emergency. These individuals include, for example, hotel managers and staff who host displaced populations in their facilities hundreds of miles away or bank employees going back when it is safe and reopening vital financial establishments in the affected community. These workers are all responding to the emergency and their response is critical to collective recovery from a catastrophic radiation emergency.

This chapter focuses mainly on professional first responders and first receivers. It describes some of their common concerns and how they can address those concerns. The following chapter talks about community preparedness.

PROFESSIONALS' CONCERNS

Opinion surveys of thousands of health care workers show a common trend. A major fraction of physicians and nurses remain apprehensive about responding to a radiation emergency and caring for radioactively contaminated patients. In fact, they would be more willing to report for work in case of a contagious epidemic than they would be after a radiological incident.* These findings, as remarkable as they may be, are not surprising. Lack of familiarity with a threat affects perception of that threat.

Physicians and nurses feel more comfortable dealing with an infectious disease because they are more knowledgeable about that threat. However, they feel uneasy with the less dangerous threat of radioactive contamination because they are not as familiar with it. The specific concerns that hospital workers have expressed in focus groups related to radiation emergencies strike a theme that is common across a spectrum of professionals.†

Professionals who may one day be asked to report to work in response to a radiation emergency are likely to have concerns about (1) safety (Will they be safe reporting to work? Will their families and loved ones be okay while they report to duty?) and (2) level of preparedness (Do they have the right tools? Do they have adequate staff? Do they know what to do in a radiation emergency?).

How workers feel about these concerns will influence their willingness to report for work in a radiation emergency and how well they perform their jobs. Understanding their organization's plan to respond to such an emergency and how they fit in, and receiving the training they need to do their jobs in that situation go a long way in addressing these concerns.

What Is Expected of You?

As a professional responder, you should have a good understanding of your roles and responsibilities in a radiation emergency. Does your organization have a specific radiation emergency response plan? If not, how does your generic, all-hazards response plan apply to a radiation emergency? To whom do you report in a radiation emergency? Who would be in charge of addressing health and safety of responders in your organization? In other words, who would be your safety officer? Would the same person serve as safety officer in a radiation emergency?

If you have a well-prepared emergency response plan for your organization, these questions are addressed in that plan. The Incident Command System (ICS) was discussed briefly in Chapter 13. An adaptation of ICS to the

---

* For example, see Qureshi, K., R. P. M. Gershon, T. Sherman, T. Straub, E. Gebbie, M. McCollum, M. J. Erwin, and S. S. Morse. 2005. Health care workers' ability and willingness to report to duty during catastrophic disasters. *Journal of Urban Health* 82 (3): 378–388; and Lanzilotti, S. S., D. Galanis, N. Leoni, and B. Craig. 2002. Hawaii medical professionals' assessment. *Hawaii Medical Journal* 61 (8): 162–173.

† For example, see Becker, S. M., and S. A. Middleton. 2008. Improving hospital preparedness for radiological terrorism: Perspectives from emergency department physicians and nurses. *Disaster Medicine and Public Health Preparedness* 2 (3): 174–184.

hospital environment is the Hospital Incident Command System (HICS).* If you work in a hospital, where do you fit in your hospital's emergency response plan? The ICS principles are more than just bureaucratic flow diagrams and lifeless terminology to memorize. They teach a tested and proven philosophy of incident management that can be adapted to any size and any type of emergency by any organization. If you work in a county health department, where do you fit in your department's emergency response plan? To whom do you report and what would be your specific responsibilities?

Keep in mind that even a well-written emergency response plan is of little or no value if it is not regularly exercised and tested. If you do not have an emergency response plan, chances are your organization will not respond to a radiation emergency or will do so ineffectively.

How Do You Do Your Job?

This question is better stated as "How do you do your job *in a radiation emergency*?" Assume that you are a well-trained individual who knows exactly what to do in your profession and does it well every day. But how do you do your job in an emergency or, more specifically, in a radiation emergency that may involve mass casualties? No one knows this better than medical professionals, who make life and death triage decisions in such situations. But consider professionals responsible for the safety of water supplies, protection of food supplies, proper sanitation, waste management, shelter hygiene, preventing or controlling epidemics, mortuary services, or crisis counseling. These professionals may feel comfortable about how they do their jobs in response to a natural disaster, but are they prepared to provide the same vital professional services in a radiation emergency? The answer to these questions is found in training.

After you have a good understanding of what could be expected of you and what your roles and responsibilities would be in a radiation emergency, you should determine your training needs. Make this assessment with the help of your supervisor and in consultation with your organization's safety officer. If training budgets are constrained, make the best use of the many free online education and training opportunities available.

Training for a radiation emergency is much more than understanding textbook concepts of radiation and radioactivity and knowing what alpha, beta, and gamma radiation is. You should know (1) what you need so that you can perform your job safely, and (2) how to perform your job to be most effective in a radiation emergency. If you have any concerns about your safety, ask

---

* See "Resources" for sources of more information about HICS.

your supervisor. Together with the safety officer in your organization, he or she should be able to address your concerns.

Teaching Family and Co-workers

The training you receive on the job for being a professional responder is useful information for your family and friends as well. You can incorporate what you learn on the job into your family emergency plan (discussed in Chapter 10). If you are a supervisor or your job involves training others, you may soon realize that communicating basic but important concepts of radiological safety can be challenging. For example, learning the difference between radiation and radioactivity is not intuitive (Chapter 3). Teaching that residual radioactive contamination does not represent an immediate health risk or that it can be simply washed off is somewhat challenging. You may encounter a look of disbelief from some members of your audience.

You can practice teaching some of these concepts to your family or close friends and gauge your ability to communicate these important messages convincingly to a lay audience. Your family and close friends are more likely to be open and share with you the points they did not understand than your co-workers may be in a formal teaching environment.

ALARA AND RADIATION PROTECTION PHILOSOPHY

Chapter 9 explored basic radiation protection principles of time, distance, and shielding. You learned what each means and how to apply these principles for protection. As a professional responder, you should also be familiar with a terminology that forms the backbone of every radiation safety program and is used whenever it is expected that some responders will be exposed to radiation as part of their job functions. The term is ALARA, which stands for "as low as reasonably achievable." The idea is that you should make every reasonable effort to reduce your exposure to ionizing radiation. In everyday uses of radiation, such as in medicine and industry, the radiation protection philosophy is that every radiation exposure should be justified and should have a net benefit for the individual. For example, a diagnostic CT scan often provides a benefit to the patient.

> *Did you know?* The x-ray technology that guides cardiologists in cardiac catheterization (coronary angiogram), angioplasty, and stent placement procedures is called fluoroscopy. The same technology was used decades ago in shoe fluoroscopes to fit shoe sizes! Shoe fitting fluoroscopes were banned in the 1950s. Clearly, fitting shoes is not a justifiable reason to

expose feet to what amounted to large doses of radiation, especially for children (Figure 14.1). Saving a life from coronary heart disease is a different matter, though.

In addition to being a justified use, every practice using radiation should be optimized. For example, if the same quality of diagnostic information can be obtained by delivering a lower dose of radiation, then CT machine parameters (such as tube current or peak voltage) should be adjusted accordingly. For pediatric patients, advances in CT technology and practice have been made to optimize and reduce radiation exposure to children while getting the same quality image as with adults.

Lastly, dose limitations are in place to limit the radiation dose each worker can receive as part of his or her routine employment. Regulations also limit the radiation dose each member of the public may receive from various industries.

Figure 14.1

Shoe-fitting fluoroscopes were a common fixture in shoe stores in the 1930s, 1940s, and 1950s. Customers would place their feet in the opening near the bottom of the wooden box, standing on top of an x-ray machine. The salesperson and the customer could then see the image of the bones of the feet and shoes' outline from viewing portals on top of the fluoroscope. This was not a justifiable use of radiation. The practice was banned in the 1950s. (Photo courtesy of Dr. Paul Frame/Oak Ridge Associated Universities, www.orau.org/ptp/museumdirectory.htm)

Did you know? In the United States, the radiation dose a member of the public may receive from various licensed operations is limited to 1 mSv (0.1 rem) per year. The radiation dose a worker in the radiation industry may receive is limited to 50 mSv or 5 rem per year, although on average, workers receive a much lower annual dose. Neither of these radiation doses results in any short- or long-term health effects.

Did you know? There are no limits on radiation doses that patients can receive at the discretion of their doctors.

The important principle of ALARA can serve the public well. The trick is not to forget the "R" in ALARA (R = "reasonable"). On occasion, tremendous societal resources are spent reducing or eliminating radiation sources with doses that are too small and trivial to be of any real consequence.

Does ALARA Apply to Radiation Emergencies?

Absolutely. It is likely that professionals responding to a radiation emergency will be exposed to some dose of radiation. This dose of radiation should be kept as low as reasonably achievable. There are specific guidelines as to how much of a radiation dose is permitted in an emergency (Table 14.1):

- If at all possible, radiation doses to emergency workers should be limited to 50 mSv (5 rem).
- For protection of valuable property or infrastructure necessary for public welfare, when no person's life is at stake, the radiation dose limit is 100 mSv (10 rem).
- For actions to save lives or protect large populations, the radiation dose limit is 250 Sv (25 rem). Some nongovernmental organizations have recommended higher radiation dose limits of 500 mSv (50 rem) or 1 Sv (100 rem) for life-saving activities and protection of large populations.*
- Any dose greater than 50 mSv (5 rem) is on a voluntary basis.
- Workers choosing to exceed recommended radiation dose guidelines on a voluntary basis should be informed of the health risks involved.

These higher radiation doses would be acceptable only if lower radiation doses are not practicable. In other words, if you cannot reasonably perform the same job incurring a lower dose, then higher doses are justified. Chances are that, for the vast majority of responders, the radiation doses they receive

* National Council on Radiation Protection and Measurements (NCRP). 2001. Management of terrorist events involving radioactive material. NCRP report no. 138. Bethesda, MD, and International Atomic Energy Agency (IAEA). 2006. *Manual for first responders to a radiological emergency.* Vienna: IAEA.

Table 14.1 Radiation Dose Guidelines for Workers in a Radiation Emergency

Radiation Dose	Activity	Conditions
50 mSv (5 rem)[a]	All occupational exposures during response to a radiation emergency	All reasonable actions are taken to minimize dose. Activities that involve radiation doses exceeding 50 mSv (5 rem) are on a voluntary basis.
100 mSv (10 rem)	Protecting valuable property or infrastructure necessary for public welfare	Workers have been fully informed of the risks of exposures they may experience. Appropriate respiratory protection and other personal protection are provided and used.
250 mSv (25 rem)[b]	Lifesaving or protection of large populations	Monitoring is available to project or measure radiation dose.

Source: U.S. Department of Homeland Security. 2008. Planning guidance for protection and recovery following radiological dispersal device (RDD) and improvised nuclear device (IND) incidents. 73 FR 45029, and U.S. Environmental Protection Agency. 1992. *Manual of Protective Action Guides and Protective Actions for Nuclear Incidents* (update available from www.epa.gov/radiation/rert/pags.html).

[a] This dose is equal to the annual limit for occupational exposure in the United States. However, in an emergency, it applies to an acute exposure.

[b] It is unlikely that radiation doses would reach this level in a radiation emergency. In the case of a catastrophic incident, such as a nuclear detonation, workers' doses higher than 25 rem (0.25 Sv) are conceivable.

in a radiation emergency (excluding a nuclear detonation) would be far less than recommended limits.*

Did you know? The natural baseline risk of getting cancer at some point in a lifetime is an average of 42%. This likelihood is in the absence of any radiation exposure. Best estimates indicate that a radiation dose of 0.05, 0.1, or 0.25 Sv (5, 10, or 25 rem) will increase the risk of getting cancer to 42.5, 43, and 45%, respectively.

* For example, an article in the *New England Journal of Medicine* reported that when highly contaminated workers at the Chernobyl nuclear power plant accident were decontaminated, the medical personnel at the site received less than 10 mGy of radiation. This is equal to 10 mSv or 1 rem, which is only 20% of the annual limit for U.S. workers. See Mettler, F. A., and G. L. Voelz. 2002. Major radiation exposure—What to expect and how to respond. *New England Journal of Medicine* 346 (20): 1554–1561.

PERSONAL PROTECTION

To emphasize again, the best protection against radiation and radioactivity is to apply the basic principles of time, distance, and shielding (Chapter 9) as appropriate. As a secondary measure, physical devices can be used to reduce or eliminate exposure to radioactive materials. The type of personal protection you may need to do your job in response to a radiation emergency depends on your responsibilities. If you work in a potentially contaminated area or have to interact physically with potentially contaminated people, you would need some form of personal protection. You would also need training on how to use that protective device appropriately—for example, learn how to put on a face mask or how to take off a contaminated pair of gloves.

Your responsibilities and the personal protection you need should be addressed in your organization's emergency response plan and procedures, and you need to receive that training *before* you are asked to respond to an emergency. This section provides an overview and some general recommendations on what protection you are likely to need depending on your job.

What Is Personal Protective Equipment?

Any specialized piece of equipment or clothing that is worn to protect against a hazard is called personal protective equipment (PPE). Many such devices are used every day by a variety of professionals working in diverse environments such as hospital emergency rooms or construction sites. Generally speaking, personal protective equipment may include items such as steel-toed boots, gloves, hard hats, safety goggles, gowns, surgical masks, or air-purifying respirators that filter the air as you breathe it. More sophisticated equipment may include a self-contained breathing apparatus (SCBA), which provides compressed air from a tank and is used routinely by firefighters (Figure 14.2).

Breathing apparatuses or face masks are effective in occupations where people have to work in environments that expose them to high concentrations of airborne contaminants; the compressed air or air filters protect them from breathing in those particles. Workers responding to hazardous material incidents may wear what is informally referred as a "moon suit" to prevent any hazardous chemicals, particles, or gas from entering the suit. The workers breathe supplied air from a tank or a supply line connected to their suits. First receivers who may receive contaminated patients may use a powered air-purifying respirator (PAPR), which takes the ambient air and filters it before the worker breathes the air (Figure 14.3).

In an incident, once the types of contaminants and their concentrations are better characterized, emergency workers may scale down to less

Figure 14.2

Firefighters in bunker gear wearing self-contained breathing apparatus (SCBA), which provides compressed air from a tank through a face mask. This personal protective equipment does not offer any protection against penetrating gamma radiation.

restrictive protective gear as appropriate. As you will see next, when radioactive materials are the only hazardous material present, most workers will not need a moon suit, or anything close to it, for performing their duties safely in a radiation emergency.

> *Did you know?* The word "scuba" (as in scuba diving) is an acronym that stands for "self-contained underwater breathing apparatus." It is the same equipment that firefighters use but is designed for underwater use.

Workers Responding to the "Hot" Zone At the site of a radiation incident, an area will be designated as the exclusion zone that only trained and properly equipped responders may enter. This area is sometimes referred to as the "hot" zone. If your work may require you to enter a hot zone, you must go through rigorous training, which includes proper selection, maintenance, and use of PPE. The site safety officer will determine what level of PPE you will need. You should know how to use the equipment properly and you must know its limitations.

The bunker gear (or turnout gear) that firefighters wear includes flame-retardant pants and overcoat, helmet, gloves, footwear, and hood

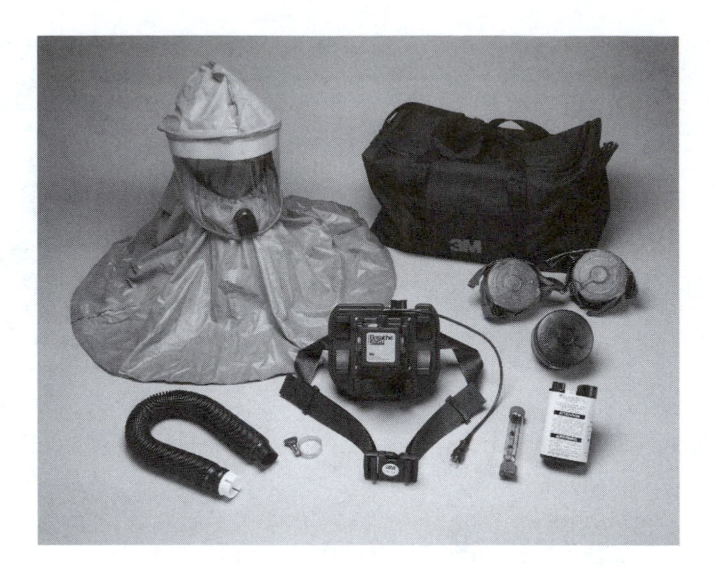

Figure 14.3

An example of a powered, air-purifying respirator (PAPR) for use by emergency responders in atmospheres that contain chemical, biological, radiological, and nuclear (CBRN) respiratory hazards. The particular type shown here does not require fit testing. (Photo courtesy of 3M.) For a list of respirators certified for CBRN environments, see the NIOSH Web site (www.cdc.gov/niosh/npptl/topics/respirators/).

(Figure 14.2). This gear will protect emergency responders from radioactive contamination getting on their skin and will stop alpha- and most beta-particle radiation. In addition, the firefighters' SCBAs will protect against radioactive contamination getting inside their bodies.

However, neither the turn-out gear nor the SCBA can protect the first responders from gamma radiation.* Even a moon suit, which provides excellent protection against inhalation hazards, offers no protection against penetrating gamma radiation (Chapter 3). This does not mean the responders should not enter an area where gamma radiation is present. Rather, it underlines the need to monitor radiation levels at a radioactively contaminated site and to provide radiation-specific training to the first responders. When engaged in lifesaving and other critical operations in a radiation field, first

* See Occupational Safety and Health Administration (OSHA) Interim Guidance on Personal Protective Equipment Selection Matrix for Emergency Responders, Radiological Dispersal Device (RDD). August 30, 2004 (available from www.osha.gov/SLTC/emergencypreparedness/cbrnmatrix/radiological.html).

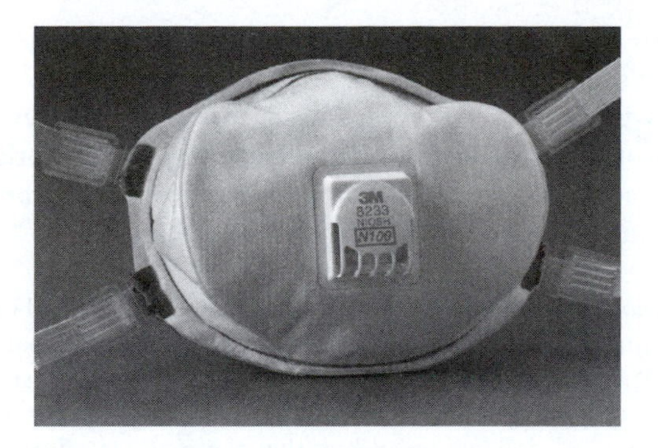

Figure 14.4
An example of an N100 disposable particulate respirator. (Photo courtesy of 3M.) For a list of NIOSH-approved particulate respirators see the NIOSH Web site (www.cdc.gov/niosh/npptl/topics/respirators/disp_part/).

responders should be aware of radiation levels and limit the time they spend in that field as appropriate.

Hospital Workers If it is known that particulate radioactive contamination is the only hazardous material present, medical personnel caring for contaminated patients can wear protective clothing and gloves as they normally do in accordance with standard precautions.* Medical personnel may also need to wear appropriate filtering face masks (such as N95 or 100 masks) to protect against inhaling radioactive particles (Figure 14.4). This determination is made after evaluating the potential inhalation hazards by a competent safety officer or radiation protection specialist (health physicist).

In the United States, the National Institute for Occupational Safety and Health (NIOSH) sets the standards for respiratory protection, and it is important to use respirators that are NIOSH approved. Hospital workers use N95 respirators for protection against hepatitis, HIV, and SARS, for example. The

* There is a small chance that a victim from a dirty bomb incident may have embedded radioactive shrapnel inside his or her body. Such a condition can be easily detected using a portable radiation detector. If radioactive shrapnel is present, it should be removed as soon as possible and placed inside a lead container or stored away.

N95 (or N100) masks can also be used in a radiation emergency if deemed necessary by a safety officer.*

Respirators work only if they are used correctly. The key elements for respiratory protection are fit-testing and training of each worker in the use, maintenance, and care of the respirator. Using a respirator *correctly* is more important than whether it has an N95 or N100 rating.

> *Did you know?* The N95 designation is an efficiency rating and means that when the mask is worn properly, it blocks at least 95% of the particles that are 0.3 microns in size or larger. An N100 mask filters at least 99.97% of such airborne particles. For first responders, the P100 masks are recommended. The "P" designation means that the mask is oil proof; this is important because some industrial oils can degrade filter performance.

Public Health and Other Workers When working closely with people who may be contaminated with radioactive materials, it is best to use booties, disposable gowns, and disposable gloves to control the spread of contamination and to wear N100 or N95 masks to protect against airborne radioactive particles. Follow directions from your safety officer as to what level of personal protective equipment you may need. For many job functions, disposable gloves may be the only required protection. In all cases, at every break period and before you leave work at the end of your shift, a radiation protection technician should monitor you to make sure you do not have residual contamination on you or your clothing. As always, using basic hygiene practices (e.g., washing your hands before eating and not rubbing your face) goes a long way to protect you from incidental contamination.

If you are ever in a situation where you realize or suspect that you may have inhaled some amount of radioactivity, remember the following: It is unlikely that your life would be in danger or that you would notice even minor health effects from small residual amounts of internal contamination.

---

* Surgical masks are not designed for use as particulate respirators and do not effectively filter small particles from air. In addition, most surgical masks do not prevent leakage around the edge of the mask when the user inhales and they cannot be fit-tested. Therefore, they do not provide as much protection as a particulate respirator such as N95 or N100.

Resources

The Federal Emergency management Agency (FEMA) Emergency Management Institute offers free online courses on a variety of topics as part of an independent study program. You can take the course interactively online or download the course material. Both options are free of charge and open to all (http://training.fema.gov/IS/). One of the courses specifically addressing radiation emergency response is IS-301 Radiological Emergency Response (http://training.fema.gov/EMIWeb/IS/is301lst.asp).

The U.S. Environmental Protection Agency's *Manual of Protective Action Guides and Protective Actions for Nuclear Incidents* is an informative document that describes radiation dose recommendations that trigger protective actions and provides useful background information. The document is being updated, but the current version is available (www.epa.gov/radiation/rert/pags.html).

For detailed information about HICS, see:

The Center for HICS Education and Training (www.hicscenter.org/pages/index.php)

California Emergency Medical Services Authority (www.emsa.ca.gov/HICS/default.asp)

For more information on the use of PPE and applicable regulations and guidance, see the following sources:

The U.S. Department of Labor, Occupational Safety and Health Administration (OSHA) has a number of pages on its Web site with general information on PPE and information specific for radiation incidents:

www.osha.gov/SLTC/emergencypreparedness/cbrnmatrix/radiological.html

www.osha.gov/OshDoc/data_General_Facts/ppe-factsheet.pdf

www.osha.gov/SLTC/personalprotectiveequipment/index.html

The U.S. Department of Labor. 2005. OSHA best practices for hospital-based first receivers of victims from mass casualty incidents involving the release of hazardous substances. Available from www.osha.gov

The CDC has information specifically related to the selection and use of PPE in health care settings:

www.cdc.gov/ncidod/dhqp/ppe.html

www.cdc.gov/niosh/npptl/topics/respirators/

The following four documents contain information that is useful for professionals planning for or responding to a radiation emergency.

Some of the information in these documents is technical and intended for radiation professionals:

Homeland Security Council. 2009. Planning guidance for response to a nuclear detonation: First edition. Available from www.hsdl.org

International Atomic Energy Agency (IAEA). 2006. *Manual for first responders to a radiological emergency.* Vienna: IAEA.

The Conference of Radiation Control Program Directors (CRCPD). 2006. *Handbook for responding to a radiological dispersal device: First responder's guide—The first 12 hours.* Frankfurt, KY: CRCPD.

National Council on Radiation Protection and Measurements (NCRP). 2005. Key elements of preparing emergency responders for nuclear and radiological terrorism. NCRP commentary no. 19. Bethesda, MD.

The following handbook was prepared by an international (primarily European) group of experts to provide emergency response organizations with guidance for field operations, medical treatment, and public health response:

TMT Handbook: Triage, monitoring, and treatment of people exposed to ionising radiation following a malevolent act. 2009. Available from www.tmthandbook.org

15

RESPONDING AS A COMMUNITY

OVERVIEW

Hurricane Katrina forced nearly a million people to leave their homes. The displaced population moved to every state in the country, and nearly every county, to seek shelter or a new home. This relocation, which unfolded in only a 2-week period, was the largest displacement of Americans since the great Dust Bowl migrations of the 1930s.* Meeting the needs of this population presented a challenge to host communities, especially in Louisiana, Mississippi, and the neighboring states that accommodated the majority of the evacuees. Public health and medical needs, housing, public safety, and social and psychosocial services were among the issues that each community had to address for the evacuees they were hosting. The resources of many communities were strained in response to this influx of evacuees.

The scope and magnitude of these disaster response issues, as daunting as they were after Hurricane Katrina, would be dwarfed by a nuclear detonation incident anywhere in the country. Depending on location of the detonation and direction of nuclear fallout, potentially several million people could be displaced after a nuclear detonation. Many among this population would suffer from radiation sickness, many people would be wounded or need some form of medical care, most would need decontamination, and nearly all would need shelter. Government agencies cannot work alone to manage

* For more information on the scope of the Katrina disaster and the government's response, see the testimony of Michael Chertoff, secretary of the Department of Homeland Security, before the Senate Homeland Security and Governmental Affairs Committee, 109th Congress, 2nd session, on February 15, 2006. For more information about the mass exodus from the American Plains in the first half of the 1930s, see Worster, D. 1979. *Dust Bowl: The Southern Plains in the 1930s*. New York: Oxford University Press.

such catastrophes. This is especially true for local governments, which carry most of the burden but have the fewest resources.

This chapter explores how the private sector, nongovernmental agencies, and individual community volunteers can help their *local* communities deal with any type of emergency where constrained resources become a limiting factor. These resource-constrained situations may involve national emergencies such as need for mass dispensing of medication to thousands of people in a community or caring for evacuees coming to a community from an area devastated by a nuclear event. There are also a number of less catastrophic and more likely scenarios, such as searching for missing persons or providing survivor assistance after a natural disaster, when local agencies can use the assistance of trained community volunteers.

The spirit of community volunteerism in the immediate aftermath of disasters appears to be a cross-cultural human trait. This spontaneous volunteerism has been demonstrated time and again around the world in response to floods, earthquakes, and other natural disasters as well as terrorism incidents. However, large numbers of volunteers who emerge spontaneously after a disaster can hinder response efforts and also endanger their own lives. The resiliency of every community depends on how well volunteer citizens and the private sector are organized and trained *in advance* and how well their efforts are coordinated and integrated with those of the local government agencies.

Two specific volunteer programs in the United States are described in this chapter: the Medical Reserve Corps and the Community Emergency Response Team. These established, locally administered programs provide the organizational infrastructure through which all individual citizens, regardless of any particular expertise or prior emergency response experience, can volunteer to serve the local community in times of need.

PRIVATE SECTOR AND NONGOVERNMENTAL ORGANIZATIONS

In the United States, the private sector owns, operates, and maintains a significant fraction of the country's critical infrastructure, including hospitals, communication networks, utilities, and financial institutions.* Effective response and recovery from major disasters requires partnership and cooperation with the private sector at all levels of government. At a fundamental level, private

* "Critical infrastructures" are assets, systems, networks, or functions so vital that their damage or incapacitation would have a significant impact on public health and safety, national security, or economic stability.

sector resiliency and plans for quick resumption of services and continuity of operation in case of a direct impact are crucial to the country's security.

In addition, the private sector can provide valuable resources during an emergency, including specialized teams, essential services, equipment, or advanced technologies. These resources, which may be donated or compensated, can be provided through local public–private partnerships or mutual aid and assistance agreements, or in response to direct requests from government and nongovernmental entities. In either case, it is best to develop contingency plans and decide on mechanisms for requesting and providing these resources ahead of time.

These partnerships are not limited to key infrastructures. Many other businesses, including retail and discount stores, should be regarded as local community partners in emergency planning. For example, in an incident where large numbers of people may be contaminated with radioactive materials and need assistance, discount stores can use the inventory in their distribution centers to provide the authorities quickly with decontamination supplies and clean clothing for people.

Nongovernmental organizations (NGOs), such as the American Red Cross, faith-based institutions, and other nonprofit or volunteer community organizations, also provide vital services in times of disaster. NGOs can provide sheltering, emergency food supplies, counseling services, and other vital support services, such as specialized services to help individuals with special needs. The success of such efforts depends, again, on how well local and state governments plan ahead and coordinate with these organizations in their communities.

Examples from Hurricane Katrina

The outpouring of support and services from nongovernmental, faith-based, and volunteer organizations in response to Hurricane Katrina was extraordinary. Similarly, private sector organizations provided commodities, services, expert advice, and financial donations. Government plans that existed at the time for incorporating and making best use of these services were less than adequate. Nevertheless, the assistance from these organizations made a significant impact on the lives and well-being of the people affected by the hurricane. A few examples include:

- In Harris County, Texas, the Citizen Corps, a volunteer organization, mobilized and coordinated thousands of volunteers to staff evacuation centers throughout Houston and to support the American Red Cross efforts. As a joint effort, the Citizen Corps volunteers helped create an actual working city for Louisiana evacuees (with its own

Figure 15.1
Citizen Corps volunteers in Harris County, Texas, helped create a shelter inside the Houston Astrodome to assist thousands of Louisiana residents who had to leave their homes because of Hurricane Katrina, September 1, 2005. (Photo: Andrea Booher/FEMA.)

zip code). It sheltered 15,000 Louisiana residents in the Houston Reliant Astrodome and thousands more in the nearby Reliant Arena, Reliant Center, and George Brown Convention Center (Figure 15.1). In all, Citizen Corps volunteers assisted in greeting and processing 65,000 evacuees from the New Orleans area.*

- An estimated 9,000 Southern Baptist disaster relief volunteers from 41 states served in Texas, Louisiana, Mississippi, Alabama, and Georgia and provided services such as mass feeding via mobile kitchens, serving 10 million prepared meals (Figure 15.2). They also assisted with a variety of other activities such as cleanup, child care, laundry, repairs, and establishing temporary shelters.†

* "Harris County, Texas, Citizen Corps' response to Hurricane Katrina," available from U.S. Department of Homeland Security, lessons learned information sharing Web site www.llis.gov (requires membership) or the Citizen Corps Web site (www.citizencorps.gov/pdf/llis/lessons-learned-tx-katrina-response.pdf).

† More information on these and other disaster relief efforts is available from the North American Mission Board news releases and annual updates (www.namb.net).

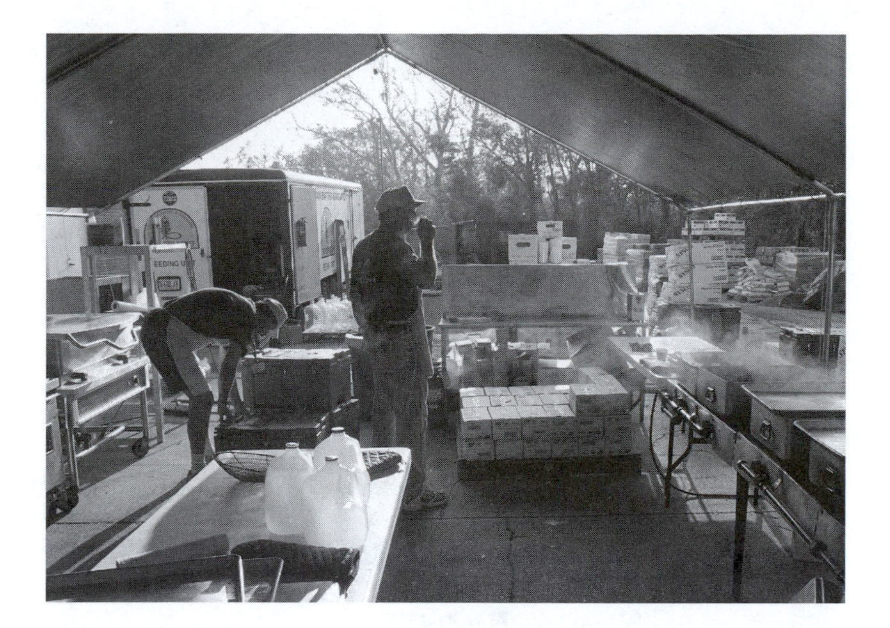

Figure 15.2
A Southern Baptist Disaster Relief volunteer prepares meals along with other volunteers from a Georgia Baptist feeding unit established in Pascagoula, Mississippi, September 15, 2005. In the aftermath of Hurricanes Katrina, Rita, and Wilma, an estimated 9,000 Southern Baptist disaster relief volunteers assisted by serving more than 10 million hot meals, participating in cleanup and recovery, providing child care and laundry services, and establishing temporary shelters. (Photo courtesy of the North American Mission Board, SBC © 2005.)

- The Salvation Army responded to immediate needs of the hurricane survivors and evacuees by serving 5.6 million meals, sheltering more than 31,000 people in seven states, and providing aid to displaced citizens in 30 states.*
- FedEx facilitated equipment and supply distribution, particularly for the American Red Cross. Dell, Home Depot, IBM, Lenovo, Pfizer, Wal-Mart, and other corporations gave millions of dollars in

* The Salvation Army disaster response history, updated August 31, 2008 (www.salvationarmyusa.org), and The federal response to Hurricane Katrina: Lessons learned, February 2006 (available from www.whitehouse.gov/reports/katrina-lessons-learned).

Figure 15.3
The Pete Maravich Assembly Center, home of the Louisiana State University Tigers and Lady Tigers basketball teams, was transformed into a field medical triage facility and then into a field and a special-needs hospital in response to Hurricane Katrina. (Photo courtesy of Louisiana State University.)

cash and in-kind donations to support immediate relief and recovery efforts as well as long-term rebuilding.*

- On the campus of Louisiana State University, a sports arena (Pete Maravich Assembly Center) and an indoor track and field facility (Carl Maddox Fieldhouse) were transformed into a field medical triage facility and then into a field and special-needs hospital. This 800-bed facility became the largest acute-care field hospital ever created in U.S. history (Figure 15.3). Volunteer students, faculty, and staff from the university provided support for these operations and also staffed a call center fielding thousands of phone calls.†

* Primary source of this information is from each company's news releases as referenced in The federal response to Hurricane Katrina: Lessons learned, February 2006 (available from www.whitehouse.gov/reports/katrina-lessons-learned).

† "Making it happen: LSU students, faculty and staff assist with relief efforts," LSU Highlights (www.lsu.edu/highlights/053/katrina.html).

In many other examples, nongovernmental and volunteer organizations as well as the private sector made a difference. It is difficult to imagine how government agencies could possibly support disaster relief activities without the support and assistance of these organizations.

INDIVIDUAL VOLUNTEERS

Even though their daily lives continue to demand more of their time, many people still appreciate opportunities to volunteer their time for a good cause, and they do so for different reasons. Some may volunteer because they like to give back to society, share their abilities, have a sense of accomplishment, or receive recognition, or because they may feel a sense of moral or religious obligation. Whatever the reason, communities certainly benefit from individual citizen volunteers who are interested in improving the community's preparedness for emergencies and other times of need.

Use of trained volunteers in emergency preparedness and response provides distinct advantages. Local and state emergency response agencies can provide services more cost effectively because it is impossible to maintain an adequate number of paid staff on a permanent basis just to cover contingencies. In an emergency, trained volunteers can increase the effectiveness of paid professional staff members by enabling them to focus their efforts on more urgent needs where they are most needed.

Furthermore, local volunteers with varied professional backgrounds can provide a broader range of expertise and experience to the local emergency response community than would otherwise be available. For example, how many people with a working knowledge of radiation safety are employed by your local health department or the county emergency management agency? In most communities, the answer is zero. However, tens of thousands of radiation specialists throughout the country work in environmental, medical, consulting, power generation, and other industries, as well as in academia. This pool of radiation experts can serve as volunteers in their communities and be available if needed.

Let us briefly explore various types of potential volunteers with respect to their prior training and professional backgrounds:

- Volunteers with medical or public health expertise include physicians, nurses, emergency medical technicians, mental health professionals, pharmacists, veterinarians, dentists, epidemiologists, toxicologists, and health educators and communicators. Most professionals in this group require active licensure and certification for practice, and some may be retired. Advanced registration of these professionals allows a volunteer agency to verify these credentials ahead of time and also address liability insurance issues as needed.

- Volunteers with other skills may include bilingual individuals, information technology specialists, computer technicians, clergy, attorneys, people with clerical skills, building inspectors, drivers, amateur radio operators, and many others. In case of a radiation emergency, radiation safety professionals (health physicists) and radiation technologists are among the desired specialties to have available locally. Some individuals in this category may also need some form of licensure for their practice. Just as was the case with medical professionals, the licensure of these volunteers may need to be verified and similar legal and liability protection issues apply.
- Trained general volunteers can support a variety of functions, some of which may require brief, just-in-time training. Examples include registering evacuees reporting to a shelter or to a reception center, or managing spontaneous volunteers, providing them with just-in-time training as needed, and supervising their activities.
- Spontaneous unaffiliated volunteers (SUVs) are individuals who, regardless of any skills they may have, have not previously registered with a volunteer agency and are therefore "unaffiliated." Generally, managing SUVs in response to an unfolding disaster can be problematic for emergency managers. It is difficult to verify volunteers' credentials quickly, they may not be covered under liability protection laws, and most have not had proper training. Advanced registration of these volunteers allows a routine background check (for criminal records) if needed, verification of their licensure or certification if applicable, and basic training on the organizational response structure in that jurisdiction.

Volunteer Organizations

Faith-based organizations have traditionally provided a venue for volunteers to participate in disaster response. These organizations usually partner with other response agencies to provide shelter, food, housing, and other disaster-related services. Faith-based organizations provide assistance to their local communities and neighboring jurisdictions, and some volunteer organizations provide disaster assistance worldwide. The focus in this section is on the types of organizations formed specifically to strengthen the health and safety of their *local* communities and to assist in emergency response activities by their *local* government agencies.

The Citizen Corps program in the United States is an example of a successful nationwide effort to encourage citizens to participate in community preparedness. At the same time, local and state governments are encouraged to reach out to their citizens and support these volunteer efforts, provide them with needed training, and include them in their emergency response plans and exercises. An abundance of information about the Citizen Corps and affiliated programs is available (see "Resources").

Two specific Citizen Corps programs that should be of interest—the Medical Reserve Corps and the Community Emergency Response Team—are described here. If you are interested in more information about these programs, see the "Resources" section at the end of this chapter. If you do not live in the United States, these examples may provide a model for organizing similar local volunteer organizations where you live.

Medical Reserve Corps The Medical Reserve Corps (MRC) is a national network of community-based volunteer organizations. The majority of MRC units are associated or closely partnered with local health departments.* MRC volunteers receive training in incident command as well as specialized training specific to their locality. For example, if a community is prone to wildfires, winter storms, or earthquakes, volunteers receive special training regarding these particular hazards. In addition, MRC volunteers receive awareness training on CBRNE (chemical, biological, radiological, nuclear, or explosives) threats. There are a number of important features of the MRC program to consider:

- MRC units provide an organizational structure for managing and using volunteers and addressing legal liability issues.
- Contrary to what the name implies, many volunteers are not medical professionals.
- Volunteers are organized locally and utilized locally.
- When applicable, professional licensure or certifications are verified in advance.
- Member volunteers are trained and prepared in advance.
- MRC volunteers are *not* first responders.
- MRC units operate within existing local emergency response plans.
- Volunteers respond to natural disasters, terrorism incidents, or other urgent public health needs. Trained volunteers, depending on their skills, experience, and qualifications, will assist the local emergency responders in any number of functions:
    - medical care, mass dispensing of medication, or administer vaccines;

* As of July 2009, there were 823 local MRC units across all 50 states with a total volunteer membership of 179,327 volunteers.

- health education as part of a local public health initiative (for example, train community groups on how to respond to a flu epidemic);
- counseling for victims, families, and responders; and
- administrative, logistical, and communications support.

Community Emergency Response Team The focus of the Community Emergency Response Team (CERT) is somewhat different from that of the MRC program. CERT volunteers are trained primarily to better prepare themselves, their families, and their neighbors in response to an emergency in their neighborhoods. The volunteers are trained to recognize various types of hazards and protect themselves, provide immediate assistance to victims (when it is safe for them to do so), and organize spontaneous volunteers at the disaster site. In essence, CERT volunteers are trained to act as incident commanders until professional responders arrive at the scene.

A well-trained and organized CERT can supply valuable support to responders by providing them with critical information when they arrive to facilitate their response. The training that CERT volunteers receive proves extremely beneficial in emergency situations, when it may be several hours or even days before professional help arrives.

When requested, trained CERT volunteers can assist local emergency managers in other ways as well, such as searching for lost children, staffing emergency operations centers, and assisting the American Red Cross and other relief organizations with mass care.

*Legal Considerations**

Some legal liability issues are associated with volunteers providing services through a volunteer agency. What if actions taken by volunteers result in harm to the person being helped or to someone else? What if the volunteers themselves are physically harmed while at work? These questions have more legal significance in litigious environments such as that in the United States. Volunteer medical personnel providing health care services are particularly sensitive to these liability issues, which are a complex area of the law; interpretations of legal statutes may vary with jurisdiction.

Generally, states have some form of a Good Samaritan law intended to encourage emergency assistance by removing the threat of liability for any harm as long as the assistance is provided in good faith, rather than in a

* The information presented in this section is meant to raise awareness of legal issues that may apply to volunteers in an emergency response situation. The information is not intended to replace legal advice from counsel.

reckless or grossly negligent manner. An important legal step in the United States was the passage of the Volunteer Protection Act of 1997, which provided important but limited legal immunity for volunteers.* The act stated that no volunteers of a nonprofit organization or governmental entity should be held liable as long as:

- the volunteer is acting within the scope of the volunteer's responsibilities;
- if appropriate or required, the volunteer has been properly licensed, certified, or authorized by the appropriate authorities;
- the harm was not caused by willful or criminal misconduct, gross negligence, reckless misconduct, or a conscious, flagrant indifference to the rights or safety of the individual harmed by the volunteer; and
- the harm was not caused by the volunteer operating a motor vehicle, vessel, aircraft, or other vehicle for which the state requires operator's license and insurance.

Furthermore, the liability protection under this act does not apply if a volunteer engages in any misconduct that

- constitutes a crime of violence, a hate crime, or a sexual offense;
- violates a federal or state civil rights law; or
- takes place when the volunteer is under the influence of intoxicating alcohol or any drug.

The importance of preregistration of prospective disaster volunteers should be evident. The preregistration allows for (1) background screening to make sure there is no history of criminal misconduct, (2) verification of professional licensure if applicable, and (3) opportunity to fulfill basic volunteer training requirements. In some jurisdictions, liability protection standards require mandatory training of volunteers working for nonprofit organizations or government entities.

Is there a legal definition for "volunteer"? The Volunteer Protection Act defined a volunteer as an individual performing services for a nonprofit organization or a governmental entity who does not receive compensation (other than reasonable reimbursement or allowance for expenses actually incurred) or any other thing of value, in lieu of compensation, in excess of $500 per year.

As for protection of the volunteers themselves, some states provide compensation coverage to registered disaster services volunteers, placing them under the same on-the-job injury protection as that of their paid employees. But policies

* Public Law 105–19, June 18, 1997.

and procedures vary with organization and jurisdiction. When you consider volunteering for an organization, you should ask about any injury as well as liability protection coverage that may apply to you as a disaster or emergency volunteer.

Did you know? The parable of the Good Samaritan (Luke 10:25-27) involves a conversation between Jesus and a contemporary lawyer.

COMMUNITY VOLUNTEERISM IN RADIATION EMERGENCIES

The importance of community preparedness and resiliency and the value of individual citizen volunteers apply to any type of emergency, natural disaster, or terrorist incident. In this context, radiation emergencies are no different.

Certainly, if your community is directly affected by a nuclear detonation, your first priority as an individual is to survive and do what you must to protect yourself and your family and help your neighbors. No one expects you to do community service because the local health department and emergency management agencies would not be operational. In this case, the adjacent and more distant communities and jurisdictions can make the most significant contribution by receiving, assisting, and sheltering the massive number of evacuees from the area directly affected. If you live in one of those distant communities and it is a well-prepared community, you can make a positive difference in the lives of many.

In the case of a dirty bomb incident (see Chapter 5), the degree of local devastation as well as any radiation hazards would be far less severe, even nonexistent in most areas, and the local government agencies would most likely remain intact and functional. In this situation, active community volunteer organizations (such as MRC and CERT) can help local government agencies address the needs of the people who are directly affected, as well as the concerns of many others who require some form of assistance—services that trained and organized volunteers can readily and safely provide.

Resources

More information about the Medical Reserve Corps and Community Emergency Response Teams is available from their national Web sites with links to find the local groups where you live (www.medicalreservecorps.gov and www.citizencorps.gov/cert).

More information about the Citizen Corps program is available from www.citizencorps.gov

The Citizen Corps Guide for Local Officials provides brief descriptions of each program and suggestions for starting similar programs in your community (http://www.citizencorps.gov/pdf/council.pdf).

A free online course, "IS-244 developing and managing volunteers," is available from FEMA Emergency Management Institute (http://training.fema.gov/EMIweb/IS/IS244lst.asp).

A nationwide, nonpartisan organization that can be effective in organizing the private sector's vast resources and expertise to assist government in disaster preparedness and response is Business Executives for National Security (BENS), with regional offices and business partners across the country (www.bens.org).

The full text of the Volunteer Protection Act of 1997 passed by the 105th Congress (Public Law 105-19) can be found at www.gpoaccess.gov/plaws/105publ.html

Part Four

CONCLUSION

16

LIVING WITH THE THREAT

WILL WE EVER HAVE TO FACE a catastrophic nuclear or radiological incident that has an impact on our communities? In spite of what you may hear or read elsewhere, no one can answer this question with any degree of certainty. As human beings, we have a long history of dealing with all sorts of natural and man-made threats. We have a good track record of adapting to or overcoming these threats. As a general rule, we take steps to prevent a threat from occurring in the first place and we also make plans and prepare ourselves so that we can reduce the impact of that threat should it ever occur. The same approach will work with nuclear or radiological threats.

Measures to prevent such incidents from occurring in the first place are, for the most part, in the domain of governments and private enterprises with license to maintain and use radioactive materials. As citizens, we can only demand that our governments and industries continue to place a high priority on necessary measures to reduce such threats. These measures can include safe practices, security, and safeguards of nuclear materials everywhere; surveillance and interdiction methods; and necessary diplomatic measures and international cooperation and data sharing.

The second part of dealing with the threat is preparedness. For this aspect, we do not have to rely completely on the government; in fact, we should not. We need to strive for resiliency within our own community. Some people may think there is nothing they can do in a catastrophic radiation emergency to help themselves. The anxiety and fear of radiation can indeed be paralyzing. In fact, however, as citizens and as professionals we can do a lot in the face of a radiation emergency and a lot ahead of time to prepare ourselves and our communities. Our actions and reactions as citizens can influence the ultimate impact of a radiation emergency.

First, we can become informed and understand the nature of radiation—what it can do and what it *cannot* do. Hopefully, this book has been a positive step in that regard and the resources identified here can provide you with more information in particular areas that interest you. Second, we need to take the time to develop a family emergency plan and start putting together an emergency supply kit. This simple action is generic and can help our families in any type of emergency, including natural disasters. Third, depending on our profession, we need to know the ways in which our jobs may be impacted in a radiation emergency, what may be expected of us, and whether we need additional training. Last, but not least, we can choose to join a community volunteer organization or a faith-based organization that works with local government agencies in times of need. This rewarding volunteer experience is the best way to be involved at the community level and receive free training, as well as an effective way to increase our local community's resiliency for any type of emergency.

What priority should these preparation steps have in our lives? Many of us may have more immediate worries, which may include our employment, health of a loved one, or simply the next month's mortgage. Planning and preparing for a *future* emergency will never be an *immediate* priority for us compared to other daily and family concerns. If we wait until it becomes an immediate priority, it will probably be too late to prepare. Therefore, the priority we give to our individual and community preparedness is clearly a personal choice.

If we take the steps to be informed and be prepared, the sense of empowerment and self-reliance not only helps us and our communities in times of need, but also can help us in the daily challenges that we face.

GLOSSARY

Activity (radioactivity): The rate of radioactive decay processes in any material. This quantity is used to describe the amount of radioactive material and is given in units of bequerrel (Bq) or curie (Ci).

Acute radiation syndrome (ARS): Range of signs and symptoms that develop when the body is exposed to a high dose of radiation in a short period of time. Commonly known as *radiation sickness* or *radiation poisoning*, early symptoms include nausea, vomiting, and diarrhea, followed by hair loss and internal bleeding at higher doses. Range and severity of symptoms depend on the dose of radiation and can lead to death within several weeks or only a few hours.

ALARA (as low as reasonably achievable): A core principle in the practice of radiation protection that states that exposures to ionizing radiation should be kept as low as *reasonably* achievable. ALARA is a philosophical approach for controlling and managing radiation exposures for the public and the workforce. The key word "reasonable" implies that practical, social, and economic factors need to be taken into account in the practice of ALARA.

Alpha particles (symbolized by Greek letter α): Particulate form of radiation with very low penetrating ability. Alpha particles cannot penetrate the outer dead layer of skin and they are not a hazard as long as they are emitted outside the body. However, alpha particles present a significant health hazard if radioactive materials emitting them are inhaled, ingested, or absorbed through an open wound.

Background radiation: Refers to radiation levels and radioactivity in the natural environment. Natural background radiation consists of cosmic radiation from outer space, radiation from natural radioactive elements in soil and rocks and radon gas. The levels of natural background radiation vary depending on local geography as well as building construction.

Becquerel (Bq): International unit to measure the amount of radioactivity. One becquerel is a very small quantity. A unit of MBq (equal to 1,000,000 Bq) or other variations are commonly used.

Beta burn: Skin injury caused by exposure to high doses of beta particles. Beta burns are not thermal burns.

Beta particles (symbolized by Greek letter β): Particulate form of radiation with low penetrating ability. Beta particles can penetrate the skin and cause severe skin injury called *beta burn*, but they cannot penetrate deeper into the body. Beta particles present an internal hazard if radioactive materials emitting them are inhaled, ingested, or absorbed through an open wound.

Bioassay: Technique for assessing the amount of radioactive materials in the body. *Direct bioassay* measures the radiation coming directly from the body. *Indirect bioassay* measures the amount of radioactivity in body excretions such as urine or feces. Both methods use the measurement data to infer the amount of radioactivity in the body.

Contamination: Presence of radioactive materials (in the form of dust, dirt, or liquid) in places they should not be. When contamination is on surfaces of objects or people, it is called *external* contamination. When it is inside the human body, it is called *internal* contamination. Radiation emitted from radioactive materials (e.g., gamma rays or beta particles) does not cause contamination. Only coming in contact with the radioactive substance itself can cause contamination.

Countermeasures: Actions taken to alleviate consequences of radiation or contamination. Medical or pharmaceutical countermeasures refer to any drugs used in treatment of persons exposed to radiation or internally contaminated with radioactive materials.

Curie (Ci): Conventional unit to measure the amount of radioactivity. One curie is a relatively large quantity and equals 37 billion becquerel (Bq). The unit of millicurie (mCi; 0.001 Ci) or other variations are commonly used.

Decay: See *radioactive decay.*

Decontamination: Act of removing or washing contamination.

Detector: See *survey meter.*

Dirty bomb: An otherwise ordinary explosive device used to spread radioactive materials. A dirty bomb is not a nuclear bomb and has a far less severe impact. A dirty bomb can cause widespread radioactive contamination.

Dose: A measure of radiation energy transferred to any material such as a human body. There are many variations of radiation dose concept in technical jargon. It is the dose of radiation that ultimately affects health outcome.

Dose rate: Radiation dose delivered per unit of time (second, minute, or hour). This quantity can describe radiation levels in a particular area or for a particular activity.

Dosimeter: A device capable of measuring or recording radiation dose. Some dosimeters measure radiation dose in real time (ideal for first responders). Other dosimeters may require postprocessing to determine accumulated radiation dose.

Exposure: See *irradiation.*

Fallout: Radioactive debris that descend to the ground after a nuclear explosion.

Fission: The splitting of large, heavy atoms into smaller ones, thus releasing large amounts of energy in the process. Also called nuclear fission.

Fission product: Radioactive elements formed as a result of nuclear fission reactions. Fission products are produced after a nuclear detonation or as a result of nuclear reactions inside a nuclear power plant. Examples of fission products include cesium-137, iodine-131, and strontium-90.

Frisker: See *survey meter.*

Gamma rays (symbolized by Greek letter γ): Similar to x-rays, gamma rays are a highly penetrating form of radiation. Firefighters' bunker gear (turnout gear) offers no protection against gamma rays. Radioactive materials that emit gamma rays are a hazard outside and inside the body. Effective methods to protect against gamma rays include using lead as a shield or simply walking away from the area.

Gray (Gy): An international unit of radiation dose. One gray is a relatively large dose equal to 100 rad.

Ground shine: Refers to radiation coming from radioactive material settled on the ground.

Half-life: The time it takes for the activity of radioactive materials to reduce by one-half as a result of radioactive decay. This physical property is specific for each type of radionuclide and is also called the *physical* half-life. When radioactive material is inside the human body, it gets excreted through metabolic processes. The *biological* half-life, therefore, is the time it takes for the amount of radioactive material in a human body to reduce by one-half as a result of excretion. When the physical and biological half-lives are accounted for together, the quantity is called *effective* half-life.

Health physics: Field of science focused on protecting humans and the environment from possible hazards of radiation.

Hot: Jargon used to describe a large amount of radioactivity (hot source) or high levels of radiation in an area (hot area or hot spot). It does not refer to temperature. This word is used in relative terms and depends on the context. For example, radiation levels that may be called "hot" in an environmental setting are not necessarily considered as such in appropriate occupational settings.

Improvised nuclear device (IND): A crude, low-yield nuclear weapon similar to the atomic weapon detonated in Hiroshima.

Intake: The act or process of radioactive material entering the body by inhalation, ingestion, or absorption through an open wound.

Ionizing radiation: The type of radiation with enough energy to remove electrons from atoms, thereby producing ions. Examples include x-rays, gamma rays, alpha particles, and beta particles.

Irradiation: The result of being subject to radiation coming from a radioactive source. If radiation is coming from radioactive materials outside the body, it is called *external* exposure. If radiation is coming from radioactive materials inside the body, it is called *internal* exposure. Irradiation does not necessarily mean coming in contact with the radioactive material itself (see *contamination*).

KI: See *potassium iodide.*

Long-lived radionuclides: Radionuclides with long half-lives. See *half-life.*

Monitoring: See *radiation survey.*

Natural background radiation: See *background radiation.*

Non-ionizing radiation: Type of radiation with energies much lower than x-rays or gamma rays. Non-ionizing types of radiation include radio waves, microwaves, infrared, visible light, and ultraviolet radiation.

Nuclear incident: Incident involving detonation of a nuclear weapon. The impact of a nuclear incident is far worse than that of a *radiological incident.*

Plume: A "cloud" of radioactive materials moving through and dispersing in air. This cloud may or may not be visible. Radioactive materials in the plume may be in the form of fine particles or gas.

Potassium iodide (KI): Nonprescription drug to prevent the thyroid gland from absorbing radioactive iodine. It does not offer protection against radiation exposure and does not protect the body from absorbing other types of radioactive materials.

Radiation: Broad term that describes energy moving in the form of waves or particles. See *ionizing radiation* and *non-ionizing radiation.*

Radiation emergency: A general term referring to any type of *radiological* or *nuclear* incident.

Radiation exposure device (RED): A radioactive source placed to irradiate people intentionally without their knowledge. Such devices are not likely to cause contamination because the radioactive material is not dispersed. Other terms used to describe such devices include radiation emitting device, hidden source, or "silent" source.

Radiation sickness (radiation poisoning): See *acute radiation syndrome.*

Radioactive decay: The spontaneous process when an unstable radioactive atom undergoes a transformation and emits radiation.

Radiological: Pertaining to radiation and radioactive materials.

Radiological dispersal device (RDD): A device specifically designed to spread radioactive material and contaminate humans or the environment. The method of dispersal can be by means of explosion, liquid spray, or aerosol. A *dirty bomb* is an explosive type of RDD.

Radiological incident: An incident involving dispersion of radioactive materials or exposure of people to radiation. The impact of a radiological incident is far less than that of a *nuclear incident.*

Radiological survey: The process of measuring and monitoring for radiation. This radiological survey or monitoring activity can be done for people, equipment, buildings, or the environment, as appropriate.

Radionuclide: A radioactive element.

Rem: A unit of radiation dose still used in the United States. One rem equals 0.01 sievert. The unit of millirem (mrem) is equal to 0.001 rem and is often used to measure environmental doses of radiation.

Sealed source: See *source.*

Sheltering: Use of a structure for protection against radiation and/or airborne radioactive materials outside. *Sheltering in place,* usually recommended as an urgent action, means staying and seeking temporary shelter in whatever structure you are in at the moment (e.g., home, office, school, restaurant, etc.).

Short-lived radionuclides: Radionuclides with short half-lives. See *half-life.*

Sievert (Sv): An international unit of radiation dose. One sievert equals 100 rem and is a relatively large dose. The unit of mSv (equal to 0.001 Sv) or other variations are also used.

Source: Radioactive material emitting radiation. This could also refer to a machine that generates x-rays. Radioactive materials sealed in a capsule or bonded in solid form are called a *sealed source.* Some sources of radiation are natural in origin, such as radon.

Survey: See *radiological survey.*

Survey meter: An instrument used to measure radiation such as a Geiger–Muller (GM) detector. These instruments are typically portable, and when they are used to screen or "frisk" people for radioactive contamination, they are called "friskers."

APPENDIX A: EMERGENCY SUPPLY LIST

You may refer to the "Emergency Kit" section in Chapter 10 to see how to use this supply list. Most items in your emergency supply should have relatively long shelf lives. Store items in a cool, dry place that is easy to reach in the room that will be used as shelter (e.g., basement). Pay attention to expiration dates; use and replace your stored water and canned and dry foods every 6 months. Use a marker to write the date on stored batteries and replace them every 6 months. This list does not have to be assembled overnight. Build up this emergency supply gradually, based on your budget.

WATER

At a minimum, a 3-day supply of water per person is necessary, but a 2-week supply is preferred. Plan on storing 1 gallon of water per person per day.*

FOOD

At a minimum, a 3-day supply of nonperishable foods, but a 2-week supply is preferred. Select foods that require no refrigeration, preparation, or cooking. Avoid salty foods because they will make you thirsty. Examples of food you should consider include:

- ready-to-eat canned meats, canned fruits and vegetables, canned milk, juice, and soup;
- peanut butter, jelly, crackers, unsalted nuts, protein or fruit bars, dried fruit and trail mix, dry cereal, granola, cookies, hard candy; and
- baby formula and bottles.

* The water in your hot-water tank and water pipes may be used as an emergency water supply. See the joint publication of FEMA and the American Red Cross, Food and water in an emergency. 2004. Available from www.redcross.org/images/pdfs/preparedness/A5055.pdf.

MEDICATIONS AND MEDICAL SUPPLIES

You should have the following on hand:

- a 2-week supply of your prescription medications (ask your doctor or pharmacist about proper storage);
- copies of your prescriptions, medical insurance, or Medicare cards;
- nonprescription drugs such as aspirin or other pain reliever, antidiarrhea medication, antacid, laxative;
- diaper rash ointment if needed;
- medical supplies such as oxygen or asthma inhalers if needed;
- an extra battery for a wheelchair or hearing aid if needed;
- denture needs or supplies for contact lenses if needed;
- a first aid kit and a first aid book; and
- a medicine dropper.

OTHER ESSENTIAL ITEMS

Other essential items include:

- a battery-powered radio with extra batteries; hand-crank radios or weather radios are also options (a radio is one of the most important items in your emergency supply kit);
- flashlights with extra batteries;
- a wrench or pliers to turn off utilities if needed (if you turn off the gas utility, you should not turn it back on yourself even after the emergency has passed because the gas utility should be turned on only by a professional technician);
- a manual can opener, utility knives, scissors;
- a sleeping bag or warm blanket for each person;
- hygiene and cleaning products (toilet paper, paper towels, toothbrushes, soap and hand sanitizers, feminine supplies, moist towelettes, garbage bags);
- diapers if needed;
- a complete change of clothing per person, including a long-sleeved shirt, long pants, and sturdy shoes;
- paper cups and plates, plastic utensils, paper towels;
- pen and paper;
- books, games, puzzles or other activities for children to keep them occupied;
- plain chlorine bleach;
- local maps and a compass;

- a whistle to signal for help;
- a shovel;
- cash;
- a multipurpose (A-B-C type) fire extinguisher;
- respiratory masks—N95 disposable masks are relatively inexpensive (approximately $1 each)—at least one mask per person (a beard greatly reduces the effectiveness of these masks because a proper seal cannot be formed around the face); and
- copies of important family documents, such as identification and insurance policies, in a waterproof, portable container.

FOR PETS

You should have the following for pets:

- pet food, water, medications if needed;
- a bowl, litter box, and litter;
- a collar, leash, harness, carriers;
- a pet bed and toys; and
- a recent photo of your pets to help others identify them in case they are separated from you.

If you have a service animal, keep a copy of your service animal's license on hand.

APPENDIX B: RADIATION FROM MICROWAVES AND CELLULAR PHONES

MANY PEOPLE MAY HAVE CONCERNS ABOUT the types of radiation that we encounter more commonly in our daily lives that were not covered in the main text—in particular, radiation from microwaves and cellular phones. Before we describe these sources, let us briefly review some essential terminology. The word "radiation" is a broad term, and the main subject of this book concerns a specific type of radiation called ionizing radiation (Chapter 3). This type of radiation (such as x-rays) has enough energy to remove electrons from atoms and leave them as positively charged ions—hence, the term "ionizing." This ionizing action is the main mechanism by which ionizing radiation damages living tissue.

Many other types of radiation are encountered on a daily basis. These include radio waves, microwaves, radar, and ultraviolet (UV) radiation. Visible light is also a form of radiation. One important distinction is that none of these other types of radiation has enough energy to ionize atoms like x-rays can. Therefore, in the technical jargon, these other types of radiation are collectively referred to as "non-ionizing" radiation.

Can non-ionizing radiation damage living tissue? Yes, it can. But we should not treat all non-ionizing types of radiation as one class of radiation because they encompass such a wide range of energies and intensities, and, by extension, a wide range of effects on living tissue. In fact, the difference in energy among various types of non-ionizing radiation can be a million (10^6), billion (10^9), trillion (10^{12}), or a quadrillion (10^{15}) times. UV light is on the high-energy side of this range, and low-frequency radiation associated with electrical power lines is on the low-energy side. Microwaves and cellular phone radiation fall in the middle.

Some forms of non-ionizing radiation can damage living tissue if we are exposed to too much of it. Hazards of UV radiation are well known. This type of radiation is capable of producing photochemical reactions inside skin cells, leading to sunburn. UV damage may later lead to skin cancer. Many of our common household devices, however, emit radiation energies much lower than UV energies.

MICROWAVES

Microwave radiation has a million times less energy than UV radiation and is incapable of causing the molecular changes we just described. Microwave radiation can, however, be absorbed by water molecules and cause them to vibrate. The vibration of water molecules generates heat. That is how microwaves are used in cooking or warming food. If living tissue is exposed to high-intensity microwave radiation, the same phenomenon takes place inside the body. Water molecules inside body tissues start vibrating and temperature increases. If exposure is excessive and continues, serious tissue damage and death can follow.

A useful property of microwave radiation is that it bounces off metal surfaces. For this reason, when a kitchen microwave oven is operating, microwave radiation is reflected by the oven walls and does not radiate outside the oven. If the oven door does not seal properly, some leakage of microwave radiation may occur along the door's edge. Manufacturing standards limit the allowable amount of leakage. Any damage, buildup of dirt, or simple wear and tear can cause door seals to be less effective. If you suspect your oven door may not seal properly, it is better to avoid standing directly in front of and against the oven while it is operating, or, you can simply replace the oven.

Did you know? Microwave energy can be used in cancer treatment. In what is called "hyperthermia" treatment (or thermotherapy), microwave radiation is used to raise tissue temperatures to 45°C (113°F) to damage or kill tumor cells. The heat damages proteins or structures within the cells. Hyperthermia is almost always used in combination with other treatments such as radiation therapy or chemotherapy because it enhances their effects.*

CELLULAR PHONES

Cell phones transmit and receive radio frequencies that are in the microwave range. In general, high intensity of exposure to radiofrequencies in the range of microwaves can cause tissues to heat and result in adverse effects. For this reason, consumer devices emitting such frequencies, such as cell phones, need to meet certain standards in terms of energies they emit. Let us discuss who regulates cell phones, the possibility of health effects, and action items for a concerned consumer.

* For more information about hyperthermia treatments, see the National Cancer Institute Web site (www.cancer.gov/cancertopics/factsheet/Therapy/hyperthermia).

Who Regulates Cell Phones?

In the United States, the Federal Communications Commission (FCC), in consultation with other federal agencies such as the Food and Drug Administration (FDA), has established exposure limits for mobile devices using radio frequencies. Before cell phones are sold in the United States, they have to demonstrate that they meet these standards. Similarly, in Europe, the International Commission on Non-Ionizing Radiation Protection (ICNIRP) has recommended exposure guidelines that many European countries have adopted.

Radiation emitted from cell phones is regulated using a quantity called the specific absorption rate (SAR), which is a measure of radiation energy that is absorbed by the body while the mobile device is in use. In the United States and Canada, the SAR limits for mobile devices used by the public are (1) 1.6 watts per kilogram (W/kg) averaged over 1 gram (g) of tissue for the body or head, and (2) 4 W/kg averaged over 10 g of tissue for extremities (hands, wrists, ankles, and feet). In Europe, the SAR limits for mobile devices used by the public are (1) 2 W/kg averaged over 10 g of tissue for the body or head, and (2) 4 W/kg averaged over 10 g of tissue for extremities (hands, wrists, ankles, and feet). The limits in the United States and Canada are slightly more stringent.

Several Asian countries, such as Japan and China, have established their own SAR limits. These limits typically apply to the *maximum* energy the cell phone device can possibly emit. During normal operation, cell phones emit less energy than their stated SAR rating. The farther away they are from a cell phone tower, the more energy they emit, but the maximum output is their SAR rating.

If you live in the United States, it is possible to determine the radiation output (SAR rating) of your particular phone. Generally, this information is available on the safety and product information that comes with the phone or on the manufacturer's Web site. Alternatively, you can locate the FCC ID number for your phone and look up the information from the FCC database (www.fcc.gov/oet/fccid). The FCC ID number is shown somewhere on the case of the phone or device. Usually, you have to remove the battery pack to find this number.

For example, the author's cell phone has FCC ID number L6ARBM40GW. The highest SAR value for this phone is 1.16 W/kg averaged over 1 g when it is used at the ear and 0.43 W/kg averaged over 10 g when it is clipped on a belt. These SAR values are the highest output for this phone and both values are below the established limits.

Are There Any Health Effects?

The rapid increase in worldwide use of cellular phones has generated much interest in the possible health effects. As a result, many studies have been conducted to investigate the health effects of cell phone use, including international studies with several countries participating. Taking the totality of current scientific evidence into account, there is no convincing evidence that cell phone use causes adverse health effects. No clear link has been established between cell phone use and brain cancer or between cell phone use and other reported health effects such as sleep or memory problems, headaches, or infertility. The World Health Organization (WHO) states that "present scientific information does not indicate the need for any special precautions for use of mobile phones."*

However, some studies claim that such health effects are associated with cell phone use. All these studies have their limitations. The fact is that cell phone technology is still new. It is only in the last decade that children have started using cell phones in large numbers. France, for example, has now placed limitations on marketing of cell phones to children and the use of cell phones by children. That decision was not based on any scientific evidence.

More long-term studies are planned to investigate the possible health effects of cell phone use. Critics of present cell phone regulations (described earlier) argue that these regulatory limits are based on thermal effects of radiofrequency exposures and do not take into account "nonthermal" effects. However, the biological significance of nonthermal effects, leading to adverse health impacts in people, is questionable at the present time.

We should note that it is impossible to prove that any product or exposure is absolutely safe. Therefore, we cannot say conclusively that use of cell phones presents absolutely no health risks. If an associated health risk has not yet been determined, that risk is probably very small.

What Should a Concerned Consumer Do?

If you are concerned about possible health effects of cell phones, you can take some simple steps to reduce your or your children's exposure:

- Consider using hands-free devices or headsets. Because radiation intensity drops off dramatically with distance, using such devices keeps the cell phone away from your head or body. You can also use the speaker phone when possible.
- Limit the length of your calls.
- Alternate between your left and right ears every few minutes when you are speaking on your cell phone.

* See the fact sheet on the WHO Web site (www.who.int/mediacentre/factsheets/fs193/en/).

The most important safety practice to reduce the possible hazards of cell phones is to avoid use while driving. Distraction caused by using cell phones while driving creates an immediate increased risk of having an accident.

APPENDIX C: FOOD IRRADIATION

FOOD IRRADIATION PROTECTS US FROM FOOD-BORNE diseases and is not dangerous. However, there are common misconceptions, which usually stem from lack of familiarity with this important technology and decades of research behind it. The supplemental information presented here is meant to provide a brief overview of food irradiation, its uses and safety characteristics, and some of the criticisms that have been made of it.

THE TECHNOLOGY AND CALLS FOR ITS USE

Food-borne disease outbreaks occur frequently and may involve meat, vegetables, and other foods. Food irradiation, a valuable and proven technology, uses large doses of radiation to kill bacteria and parasites that could otherwise cause food-borne diseases. The technology can also control insects, reduce spoilage, and inhibit ripening and sprouting, thereby increasing the shelf life of fruits and vegetables. Food irradiation has been investigated thoroughly in hundreds of scientific studies performed over the last 50 years, as well as evaluated by international panels of experts during that period.* The health and safety authorities of dozens of countries have approved irradiation for a variety of food items.†

In the United States, the Food and Drug Administration (FDA) first approved the method for controlling insects in wheat and flour in 1963.‡ Since then, the FDA has approved additional uses to inhibit sprouting in white potatoes, control parasites in pork, control insects and increase shelf life in fruits and vegetables, sterilize herbs and spices, and control food-borne

* World Health Organization. 1999. High-dose irradiation: Wholesomeness of food irradiated with doses above 10 kGy. Report of a Joint FAO/IAEA/WHO Study Group. Technical report series no. 890. WHO: Geneva; World Health Organization. 1994. Safety and nutritional adequacy of irradiated food. WHO: Geneva.

† International Atomic Energy Agency (IAEA). 1999. Facts about food irradiation. Vienna: IAEA. IAEA maintains a current and searchable database on country approvals of irradiated foods for human consumption. The Food Irradiation Clearance Database includes country name, class of food, specific food product, objective of irradiation, date of approval, and recommended dose limit (http://nucleus.iaea.org).

‡ United States General Accounting Office. 2000. Food irradiation, available research indicates that benefits outweigh risks. Washington, D.C. (available from www.gao.gov).

bacteria in poultry and meat. In 2008, the FDA approved irradiation of fresh iceberg lettuce and fresh spinach.*

Irradiated foods have been endorsed as safe by many organizations, including the World Health Organization (WHO), the Food and Agriculture Organization (FAO) of the United Nations, the American Medical Association (AMA), the American Dietetic Association (ADA), the Centers for Disease Control and Prevention (CDC), the U.S. Department of Agriculture (USDA), and the FDA.

Some key points about food irradiation include:

- Irradiation does not make food radioactive.
- Irradiation at approved doses kills most bacteria (e.g., *Escherichia coli, Campylobacter, Salmonella, Listeria*) and parasites in food; however, currently approved radiation doses do not kill viruses or protect against botulism.†
- Irradiation does not change or diminish the nutritional value of foods. A possible small loss of nutrients during irradiation is similar to that of other food processing methods, such as pressure cooking, canning, and heat pasteurization.
- Irradiation produces minor chemical changes in foods, as do cooking and barbecuing, but irradiated foods are not harmful or dangerous.
- Irradiation is a cold process and does not cook the foods.

CRITICISMS OF FOOD IRRADIATION

Critics of food irradiation assert that it has the potential to make food manufacturing and processing companies less attentive to standard food safety practices and too reliant on radiation to eliminate harmful pathogens in their products. It is important for food processing facilities to recognize that food irradiation is not a panacea and that it does *not* eliminate or reduce the need to operate comprehensive food safety and sanitation programs.

It is always advisable to prevent food contamination in the first place. In fact, food irradiation is not recommended for foods already known to be contaminated. As noted earlier, food irradiation cannot eliminate all food pathogens and does not eliminate any chemical contaminants. All of us should realize that buying an irradiated food product does not mean that we can forget about using proper food handling procedures at home. Irradiated

---

* Irradiation in the production, processing and handling of food (73 FR 49593), August 22, 2008. Also see the FDA's Consumer Health Information Web page (www.fda.gov/consumer).

† Higher radiation doses are necessary to kill *Clostridium botulinum*.

foods that we buy can still become contaminated at home if we do not use proper procedures.

Some critics assert that food irradiation alters food and introduces hazardous byproducts. These claims are not supported by the extensive body of scientific research conducted over the last few decades and the consensus of scientific opinion that is based on that research.*

DID YOU KNOW?

When Did Food Irradiation Begin?

The first patent was issued in Britain in 1905 to use radium as a tool for food preservation.† U.S. patents were first issued in 1918 and 1921 to use x-rays for food preservation and to control parasitic disease in pork meat.‡ It took decades for these types of applications to become viable on a large commercial scale. The National Aeronautics and Space Administration (NASA) supplied astronauts with irradiated food beginning in the 1970s.

What Sources of Radiation Are Used to Irradiate Foods?

Three types of ionizing radiation can be used for food irradiation: gamma rays, x-rays, and high-energy electron beams. The most common type of irradiation machine is a gamma-ray industrial irradiator that uses cobalt-60 as the radioactive source.

What Radiation Dose Is Necessary to Irradiate Foods?

Specific protocols are approved for each application by the FDA. For example, the radiation dose it takes to inhibit sprout development in potatoes is less than the dose it takes to kill bacteria. Generally, radiation doses range from 0.5 kilogray (kGy) up to 10 kGy in approved protocols for food items.§ (1 kGy

* In addition to the WHO 1994 and 1999 reports referenced earlier, see Wood, O. B., and C. M. Bruhn. 2000. Position of the American Dietetic Association: Food irradiation. *Journal of the American Dietetic Association* 100 (2): 246–253, and find additional information at the Web sites of the CDC (www.cdc.gov/Ncidod/dbmd/diseaseinfo/foodirradiation.htm) and FDA (www.fda.gov/opacom/catalog/irradbro.html).

† Diehl, J. F. 1995. *Safety of irradiated foods*, 2nd ed. New York: Marcel Dekker, Inc.

‡ The 1921 patent was for controlling *Trichinella spiralis* in pork meat responsible for trichinosis.

§ For example, strawberries are frequently spoiled by *Botrytis* mold. A radiation dose of 2–3 kGy, followed by storage at 10°C (50°F), can extend their shelf life to 14 days.

Figure C.1
The Radura is the international symbol for irradiated food. The label must be accompanied with the words "treated with radiation" or "treated by irradiation." The color of the logo is green, although it can be printed in black and white. (Photo: USDA Food Safety and Inspection Service.)

= 1,000 Gy or 100,000 rads). Just for comparison, the radiation doses used in food irradiation are approximately 1,000 times the lethal dose for humans.

How Can I Identify Irradiated Foods?

In the United States, all irradiated foods must be labeled as such. The international symbol for irradiated food is the Radura, which portrays a plant inside a circle and is usually green, although labels can be printed in black and white (Figure C.1). The label must be accompanied with the words "treated with radiation" or "treated by irradiation."*

Laboratory tests can determine if food has been irradiated. One of these techniques is electron spin resonance (ESR) spectroscopy. Another test involves the detection of a potentially unique compound that has been identified in irradiated meat and poultry: 2-alkylcyclobutanone (2-ACB). When food containing fat is irradiated, very small amounts of this compound are produced that are not found in nonirradiated foods; therefore, it can be used as an indicator of food irradiation.†

* For irradiated foods not in packages (e.g., fruits and vegetables), the logo and required statement should be displayed to the purchaser by labeling the bulk container plainly in view or on a counter sign or card bearing the information. See the U.S. Code of Federal Regulations 29 CFR 179.26(c) for legal requirements.

† Although some claims of adverse health effects have been made, the large body of scientific evidence does not support any adverse health effects from 2-ACB in quantities produced in irradiated foods.

What Other Products Are Irradiated?

Some hospital supplies and medical products, as well as cosmetics, some food packaging materials, and wine bottle corks, are irradiated in industrial irradiators to sterilize them.

INDEX

J

L

M

N

R